This undergraduate text takes the reader along the trail of light from Newton's particles to Einstein's relativity. Like the best detective stories, it presents clues and encourages the reader to draw conclusions before the answers are revealed.

The first seven chapters describe how light behaves, develop Newton's particle theory, introduce waves and an electromagnetic wave theory of light, discover the photon, and culminate in the wave-particle duality. The book then goes on to develop the Special Theory of Relativity, showing how time dilation and length contraction are consequences of the two simple principles on which the theory is founded. An extensive chapter derives the equation $E=mc^2$ clearly from first principles and then explores its consequences and the misconceptions surrounding it. That most famous of issues arising from special relativity – the aging of the twins – is treated simply but compellingly.

The book grew out of a popular one-semester course for non-science students. No previous knowledge of physics is required; of mathematics, only high school algebra is needed. Experiments provide the motive and the testing ground for physical theories. Especial emphasis is therefore given to the description of pivotal experiments. Moreover, details are provided on how to perform many of the experiments as lecture demonstrations. Biographical and historical sketches complement the physics and provide a human richness that students welcome. Many exercises, constructed with an emphasis on conceptual questions, are provided.

Newton to Einstein : The trail of light

*An excursion to the wave-particle duality
and the Special Theory of Relativity*

NEWTON TO EINSTEIN
The trail of light

An excursion to the wave-particle duality
and the Special Theory of Relativity

RALPH BAIERLEIN

Wesleyan University, Middletown, Connecticut

CAMBRIDGE
UNIVERSITY PRESS

Published by the Press Syndicate of the University of Cambridge
The Pitt Building, Trumpington Street, Cambridge CB2 1RP
40 West 20th Street, New York, NY 10011–4211, USA
10 Stamford Road, Oakleigh, Victoria 3166, Australia

First published 1992

Printed in Great Britain at the University Press, Cambridge

A catalogue record of this book is available from the British Library

Library of Congress cataloging in publication data

Baierlein, Ralph.
   Newton to Einstein : the trail of light: an excursion to the
wave-particle duality and the special theory of relativity / Ralph
   Baierlein.
      p.   cm.
   ISBN 0–521–41171–8 (hardback)
   1. Wave-particle duality. 2. Special relativity (Physics)
I. Title.
QC476.W38B35 1992
   535′.1 — dc20   91–25083   CIP

ISBN 0 521 41171 8 hardback

Contents

To my students in Physics 104, Newton to Einstein. Your evident interest, the varied insights of your essays, and our shared enjoyment of the demonstrations encouraged me to write this book.

Preface

The college catalog said

> Physics 104. Newton to Einstein: The Trail of Light
>
> The course will follow the trail of light from Newton's corpuscles to Einstein's relativity. The major theoretical landmarks will be the wave-particle duality and the special theory of relativity.

It was the spring term of 1987, and the course was brand-new. Out of my previous experience with courses for non-scientists, I expected a class of 20 to 30 students. If enrollment became a fad, then the class might grow to 50 or 60. I was not prepared for what happened. When pre-registration was over, 141 students had signed up.

As an aside, let me remark that I was also scheduled to teach the calculus-based introductory course, ordinarily our largest course. That spring I taught more students than the rest of the department, all put together.

My intention was to teach the course in alternate years (to maximize the number of times I would teach it before burning out). When I taught the course the second time, in the fall of 1988, over 170 students sought admission. Reluctantly, I held the line at 150, my saturation point when reading essays and correcting exams.

During the last week of the semester, I ask my students for advice about improving the course, and I also ask them, "What, for you, is the most valuable thing you learned or 'got' from this course?" In 1988, my younger son, then a senior in high school and ever skeptical of his father's teaching methods, happened to read the responses first. When he handed them back, he remarked laconically, "I'm impressed." I thought to myself, maybe there is something in this course that others would find useful. I had not been able to find a textbook – rather, I had relied on the lectures and on photocopied notes – and so I set out to turn my notes into this book.

How, you may wonder, is the book organized? Since the book grew out of a course, it reflects a specific progress through the course's topics. But I have tried to make the book modular. Few professors want to commit themselves to following an author's development step by step; as a breed, we professors are far too independent-minded for that.

When I taught from a draft in 1990, my own sequence started with section 1.1 and then went directly to chapter 2, returning to other sections of chapter 1 as they were needed in the development. An instructor who does not want to include Newton's particle theory may omit chapter 2 entirely.

In writing the chapters on relativity theory, I deferred the twins so that students can become familiar with time dilation (through homework questions) before the text takes up the twins. Getting to the twins is what I aim for in a semester's course; the chapter on the Lorentz tranformations *per se* is there for an instructor with more time or a different set of priorities.

An instructor can even use the relativity chapters independently of the first seven chapters. To be sure, if the development of $E=mc^2$ is to be included, one must get the idea of energy and momentum for a photon from somewhere (if not from chapter 6). The general idea of energy is developed in appendix A, where it is available whatever route an instructor takes through the book. A glossary collects many of the technical terms that are defined at various points in the book.

In the first half of the book, which takes the reader through the wave-particle duality, the development is chronological and historical, broadly speaking. The development of relativity theory, however, is ahistorical. The two fundamental principles are taken as generalizations from everyday experience and from a particularly clean experiment performed in 1964. Let me explain why I adopted this approach.

The first and second times that I taught the course out of which this book grew, I asked my students to read Banesh Hoffmann's *Relativity and Its Roots*. The book is delightfully written, and I enjoy it immensely. Yet in the long, tortuous history of the ether, my students got confused by all the things – new to them – that turned out to be contradictory or wrong or irrelevant. In 1905, Einstein remarked that "the introduction of a 'luminiferous ether' will prove to be superfluous." Good advice. I took it to heart.

When I was in graduate school, the Physical Science Study Committee developed its innovative high school course. I had the privilege of teaching with Eric Rogers, one of the developers, and I absorbed a lot of the PSSC spirit. Anyone familiar with PSSC physics will recognize its influence in the first few chapters of this book and in some of the homework questions.

The book *Revolutions in Physics*, by Barry Casper and Richard Noer, was recommended to me some ten years ago. I have occasionally taught from parts of it, and I have mined it for ideas and homework questions. My acknowledgement and thanks go to Professors Casper and Noer for an excellent book.

Teaching the equation $E=mc^2$ is difficult, primarily because of the misconceptions with which students come to the topic and because there is no uniquely correct way to go about describing the deep connection between inertia and energy. Chapter 11 provides one correct way of presenting the topic; an alternative is discussed in appendix C, More about $E=mc^2$. In both places I take great pains to forestall and combat misconceptions; I welcome correspondence on how I could do the job even better.

Every instructor has his or her own way of approaching a subject and teaching a course. I have no intention of telling people how to go about their work; yet some information about how I teach my course may be useful to some instructors, and so I will say a bit more here.

My goals in the course are the following four: (1) to give my students a positive experience with science; (2) to teach them some intellectually important physics; (3) to give them experience with clear, logical thinking, both mine (in lecture and text) and theirs (in the homework); and (4) to provide some acquaintance with the human and historical sides of physics.

The syllabus tells my students that "homework is for learning, not for testing. The staff and I will *correct* your homework but not grade it. The only record we keep about each question is whether or not you made a serious attempt to answer it." My students find this policy supportive; it reduces their fear of science. And the policy enables me to pose some quite challenging questions.

My lectures and this book cover the essential physics, but the course includes other books, usually three of them. They have been Edward Andrade's *Sir Isaac Newton*, Albert Einstein's *Ideas and Opinions*, and either Banesh Hoffmann's *Relativity and Its Roots* or Richard Rhodes's *The Making of the Atomic Bomb*. Of these books, only the Einstein volume is wholly satisfactory in my context, and so I am always on the lookout for other books.

With each book, I assign a short paper (of two or three pages), a practice that has many merits. The papers give my students an opportunity to do what many of them do well, namely, to write, and thus the papers are a good way to build up credit toward a good course grade. Moreover, the essays engage the mind and provide an opportunity for self-expression in what is generally too passive a learning context. The insights into Newton and Einstein prove to be fascinating and to shatter stereotypes. The first batch of Newton papers that I read contained a sentence that I will never forget. Tim Orr wrote, "Suffice it to say that when I read the biography, I was not expecting a very colorful picture of Newton, but, as I found out, the life of the man that I learned to hate in high school is really quite interesting, and I would not mind learning even more about him."

Demonstrations are a vital part of my teaching, and that is one reason why I teach so often about light: the demonstrations work (for there is no friction!), and they are often beautiful. When I ask my students for advice about the course, one of the questions has been this: "If there is a topic or demonstration that you particularly enjoyed, tell me that, too." In 1988, a student had this to say in response: "Mostly, the demonstrations are a GREAT way of seeing what you say – without them, I don't know how I'd learn any of it or believe you." I agree, and I wouldn't know how to give a lecture without some props. Beyond that, all of us enjoy the demonstrations, and they provide something to look forward to when the alarm clock goes off for an early morning class.

The course has a mid-semester hour exam and a cumulative final exam. Throughout the course, I keep in mind the question, what will my students remember five years from now? The equations? The sense of logic? The demonstrations? The flavor of the course? My personality? It is humbling, and it leads me to do only a few things but to do them well. If anything, I still address too many topics, though by no means all that appear in this book.

Enough of organization; let me go on.

The lecture demonstrations would never have been so varied and so successful without the assistance of our physics curators, first Wlad Miglus and then his son and successor, Vacek Miglus. My thanks go to them for so much help in the prep room before the next class.

My colleagues at the university and in the physics community at large have been a great help to me. My special thanks go to Phyllis Fleming, A. P. French, Stewart Gillmor, Steven Lebergott, Richard Lindquist, Janet Morgan, Stewart Novick, Joseph Rouse, and Penny Russman. Another bouquet of thanks goes to Anne Stevenson, my secretary while the course and book emerged from lecture notes. Some of my students were particularly helpful – reading drafts or asking good questions – and so I send my thanks to Holly Adams, Kathy Booth, and David Lakein.

At the Press, Rufus Neal was encouraging and helpful – always – and to him I express my warm appreciation. Finally, I thank my wife Jean for her good advice and steadfast support.

Middletown, Connecticut　　　　　　　　　　　　　　Ralph Baierlein
March 1991

1 How light behaves

First gather the facts; then you can distort them at your leisure.

Mark Twain

1.1 First observations

How should we begin? Surely with Isaac Newton. He published his first scientific paper in 1672, and it was on light. In response to an objection raised against his paper, he wrote

> For the best and safest method of philosophizing seems to be, first diligently to investigate the properties of things and establish them by experiment, and then to seek hypotheses to explain them. For hypotheses ought to be fitted merely to explain the properties of things and not attempt to predetermine them, except in so far as they can be an aid to experiments.

So let us get a light source and begin investigating. I will suppose that you have access to some apparatus – at least in the sense of seeing lecture demonstrations – and so I will describe things as though we were doing the experiments together.

We take a strong flashlight or a 35-millimeter slide projector and shine the light horizontally. Clapping a pair of well-used erasers produces a cloud of chalk dust, and we see the light beam piercing the cloud as a bright shaft of light. This gives us our first observation:

Observation 1. Light goes in a straight line from a luminous source.

Reflection and refraction

Next we tip our source – the projector, say – so that the light enters the water in an aquarium tank. Figure 1.1 depicts what happens. The beam of light changes direction as it enters the water. This bending is called *refraction* (from the Latin "to break," as in the word "fracture"). We must augment our first observation to read as follows:

Observation 1′. Light goes in a straight line (so long as it moves through a single uniform substance).

1

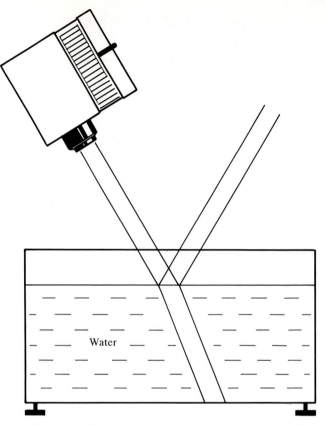

Figure 1.1 A light beam striking the surface of water: refraction and reflection. In a dark room, ordinary tap water gives a visible refracted beam. Chalk dust will reveal the reflected beam.

Before we go on with the experiment, some remarks about the glossary are in order. Many technical terms, like the word *refraction*, are collected and defined once again in the glossary, which is printed near the end of the book. Regard that collection as a valuable resource, particularly when you review. In fact, it is a good idea to take a look at the glossary right now. You will get a further sense of what is to be found there, and a visual impression will make its existence stick in your mind.

Now we return to our light source and the tank of water. In addition to the refracted beam, we note also a reflected beam of light. How can we characterize it geometrically? Figure 1.2 gives a stylized rendition of the situation; beams of light (of some thickness) have been replaced by lines and arrow heads, indicating single rays of light. The perpendicular to the surface has been drawn in; it is called the *normal* to the surface.

To the eye, the angle between the reflected ray and the normal appears

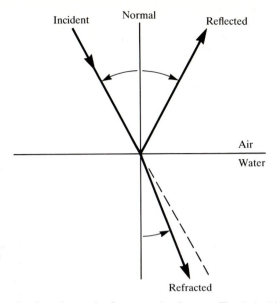

Figure 1.2 A stylized rendition of reflection and refraction. The dashed line shows the path that light would take if it were not refracted. The angle between the normal and the incident ray is called the *angle of incidence*; that between the normal and the refracted ray, the *angle of refraction*; and that between the normal and the reflected ray, the *angle of reflection*.

equal to the angle between the incident ray and the normal. Careful measurement (with more sophisticated apparatus) corroborates this. Moreover, if the incident ray was traveling due east (and downward), then the reflected ray travels due east (and upward). The horizontal motion is the same for both rays. We codify these remarks as

Observation 2.

(*a*) The angle of reflection equals the angle of incidence.

(*b*) The incident ray, the reflected ray, and the normal lie in the same plane.

As an aside, let me remark that Observation 2 provides the basis for understanding mirrors, both bathroom and funhouse.

Reversibility

Next, we can try sending light back along the path it has taken. To do that, we put a mirror in the path and adjust the orientation until the beam heads straight backward.

What happens when the returning beam encounters a place where the outward-going beam was reflected (as in the upper portion of figure 1.2)? Observation 2 tells us that light coming in along the "reflected" direction

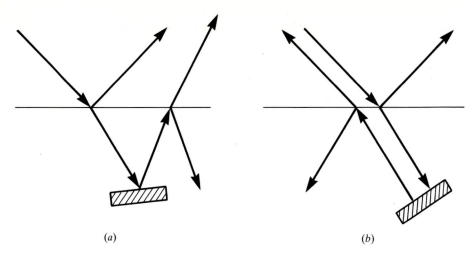

(a)   (b)

Figure 1.3 A mirror in the water tank. (a) While we are trying to find the correct
orientation for the mirror, the light beams will look something like this. (b) When we
have achieved the orientation that sends light straight backward *within* the tank, the light
goes straight backward also after refraction into air. (The arrows have been displaced
sideways for clarity's sake only.)

will be sent out along the "incident" direction. The equality of the angles
ensures this. (There may also be a new refracted beam if the reflection is not
from a mirror.)

If, however, the returning beam encounters a place where the outward-
going beam was refracted, then our observations to date do not tell us what
will happen. We need a new experiment. Figure 1.3 shows what happens if
we put a mirror into the tank of water. The newly refracted beam retraces
the original path in air (and there is an extraneous reflection back toward
the bottom of the tank).

We find that each reflection and each refraction is reversible. A light path
made up of several such occurrences, in any order, will also be reversible.
Thus our investigations lead to

Observation 3. Light paths are reversible.

To be sure, sending a beam back along its original path may lead to extra
branching, as in part (b) of figure 1.3, but a portion of the light will make it
back all the way along the original path.

More about refraction

Now we turn to characterizing the refracted beam. In figure 1.2, the perpendicular to the surface extends on both sides of the surface, and so we can
speak of an angle of refraction as the angle between the refracted ray and

Table 1.1. *Data on refraction for the transition air-to-water. Angles are measured in degrees. The column headed "Ratio of angles" gives a quotient: the angle of incidence divided by the angle of refraction. The fourth column gives a similar quotient: the semi-chord associated with the incident ray divided by that for the refracted ray.*

Angle of incidence	Angle of refraction	Ratio of angles	Ratio of semi-chords
10	7.5	1.33	1.33
20	14.9	1.34	1.33
30	22.1	1.36	1.33
40	28.9	1.38	1.33
50	35.2	1.42	1.33
60	40.6	1.48	1.33
70	45.0	1.56	1.33
80	47.8	1.68	1.33

the normal on its side of the surface. The angles in this figure and in figure 1.1 were drawn faithfully, and so inspection of the drawings gives us

> Observation 4. On the transition air-to-water, the angle of refraction is less than the angle of incidence.

In short, the beam is bent toward the normal.

How does the angle of refraction change if we change the angle of incidence? For example, if we double the angle of incidence, does the other angle double also? Table 1.1 provides some experimental data. The ratio of angles is *not* constant, and so the angle of refraction is not proportional to the angle of incidence. The ratio of angles shows a systematic trend, however, and so there is still hope for a simple relationship.

Figure 1.4 lays out the geometry more fully. An incident light ray, the corresponding refracted ray, and the normal are present. In addition, I have drawn a circle of radius R around them and have drawn in the semi-chords. The numerical value of the radius R is inconsequential, for only ratios will be important to us. The point of figure 1.4 is to introduce semi-chords into our description of refraction.

If we examine the ratio of semi-chords in table 1.1, we find that a constant ratio emerges. Thus we have

> Observation 5. In refraction, the ratio of semi-chords is the same for all angles of incidence (in the transition air-to-water).

At this point, several thoughts may arise in your mind. First, this relation-

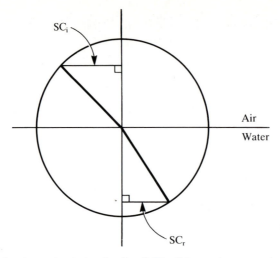

Figure 1.4 Refraction and a circle of radius R. The light ray intersects the circle at two points. The lines drawn from the points of intersection to the normal are the semi-chords. They meet the normal perpendicularly. The letters "SC" denote "semi-chord," and the subscripts "i" and "r" stand for "incident" and "refracted," respectively.

ship really isn't simple. Yes and no. It is not an intuitively obvious result, but any time we can find a direct proportionality – here between semi-chords – we have something at least relatively simple.

Second, why does it work out this way for light? An answer to this question requires some theory building; that comes in chapters 2 and 3. Right now we are still gathering data, as Newton advised us to.

Third, of what use is Observation 5? It has many uses, really. For us, it summarizes table 1.1, and we can use it to calculate refraction for angles not listed in the table. More importantly, we can test any theory of light by asking whether it reproduces Observation 5. And finally, the observation provided the original basis for the systematic design of lenses, from spectacle lenses to telescope lenses.

An illustration of how we can calculate new angles of refraction is in order. Suppose the angle of incidence is 45 degrees. What is the angle of refraction?

For the transition air-to-water, table 1.1 shows that the ratio of semi-chords is equal to 1.33. So we may write

$$\frac{\text{semi - chord of incident ray}}{\text{semi - chord of refracted ray}} = 1.33.$$

When we let "SC" stand for "semi-chord" and let subscripts "i" and "r" stand for "incident" and "refracted," the relationship takes the concise form

$$\frac{SC_i}{SC_r} = 1.33.$$

To solve for SC_r, we multiply both sides by SC_r and divide both sides by 1.33:

$$\frac{SC_i}{SC_r} \times \frac{SC_r}{1.33} = 1.33 \times \frac{SC_r}{1.33}.$$

Upon cancelling in numerator and denominator, we emerge with the equation

$$\frac{SC_i}{1.33} = SC_r.$$

Literally measuring SC_i in a faithful drawing, dividing by 1.33, and then constructing SC_r to be of the new length will generate the correct refracted ray. Figure 1.4 was actually constructed in this fashion. But shortly we will meet a more analytic way of achieving the same end.

We move on a bit. When a light beam goes from air into glass, the beam is refracted, but – once again – the ratio of semi-chords is the same for all angles of incidence. The same constancy occurs with the transition from air into any other transparent substance.

The constant value of the ratio SC_i/SC_r is called the *index of refraction* for the transition air-to-whatever-substance. For the transition air-to-water, we know the index is 1.33. What about other substances? For air-to-glass, the index is about 1.5. That means the angle of refraction is smaller relative to the angle of incidence than is true for air-to-water; the light beam is bent more. For a typical clear plastic, the index is about 1.45. Gaseous nitrogen, which constitutes 80 per cent of the air we breathe, can be cooled to form a transparent liquid. For nitrogen in that liquid state, the transition air-to-nitrogen has an index of only 1.21. The small value implies less bending than in the case of air-to-water. This diversity of numerical values leads us to make

Observation 6. The index of refraction depends on the pair of substances.

In less technical terms, Observation 6 just says that, if we specify a fixed angle of incidence, the amount of bending depends on the pair of substances.

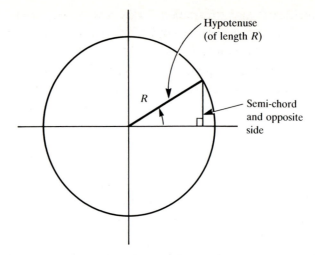

Figure 1.5 In a circle of radius R, we have a right triangle in an orientation that may be particularly familiar to you. The angle formed at the circle's center is the focus of our attention. The side opposite that angle is a semi-chord.

Now that we have gained a little familiarity with refraction, we can return to Observation 5 and phrase it somewhat differently. Figure 1.5 presents the essential triangle. For any such right triangle constructed in a circle of radius R, we have the relationship

$$\text{sine of angle} = \frac{\text{opposite side}}{\text{hypotenuse}} = \frac{\text{semi-chord}}{R}.$$

Thus the sine and the semi-chord are proportional to each other, and so a ratio of semi-chords is equal to a ratio of sines. We apply this insight to the two right triangles in figure 1.4. (Simultaneously, we confirm the mathematical reasoning.) We write down the ratio of semi-chords, divide both numerator and denominator by R, and then recognize a ratio of sines:

$$\frac{SC_i}{SC_r} = \frac{SC_i / R}{SC_r / R} = \frac{\text{sine of angle of incidence}}{\text{sine of angle of refraction}}.$$

Because sines are sometimes easier to work with than semi-chords, we rephrase Observation 5 to read as follows:

> Observation 5′. In refraction, the ratio of sines is the same for all angles of incidence (but the ratio does depend on the pair of materials).

Succinctly,

$$\frac{\sin (\text{angle of incidence})}{\sin (\text{angle of refraction})} = \text{index of refraction}.$$

In this form, the observation is called *Snell's law*, an empirical relationship formulated by Willebrord Snell around 1621. A professor at the University of Leiden in the Netherlands, Snell taught mathematics, astronomy, and optics. Although Snell had spent years of experimentation and thought in arriving at his law of refraction, he did not publish the result in the five years that remained of his life. The law first appeared in print in Rene Descartes's *Dioptrique*, almost two decades after Snell discovered it.

Total reflection

The next observations are most easily made with a thick semi-circular piece of plastic or glass, rather than with water. Figure 1.6 sets the context. A light beam in air strikes the center of the plastic's diameter and is refracted into the semi-circular material. (There is no further bending when the beam leaves the plastic and re-enters air because the beam hits the second surface dead on.)

When we increase the angle of incidence, the angle of refraction grows. As the angle of incidence approaches 90 degrees, the angle of refraction increases to some maximum angle that is less than 90 degrees.

Indeed, we can determine that angle from figure 1.6 and the angle of incidence. The semi-circle in the lower half of the figure denotes the curved boundary of the plastic, but – in your mind's eye – you can continue the arc around into the upper half to give us a circle of radius R. Then we can talk

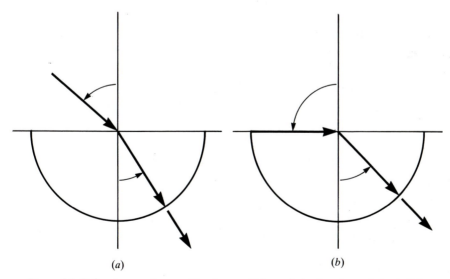

(a)   (b)

Figure 1.6 Refraction on the transition into a thick semi-circular piece of plastic (lying below the horizontal line). (a) Angle of incidence equals 50 degrees. (b) Angle of incidence approaching 90 degrees.

about semi-chords for both the incident and refracted ray. If the index for the plastic is 1.45, we have, in general,

$$\frac{SC_i}{SC_r} = 1.45.$$

As we did earlier, we can solve for SC_r and find

$$SC_r = \frac{SC_i}{1.45}.$$

As the angle of incidence approaches 90 degrees, SC_i approaches the radius R of the circle, and so

$$SC_r \text{ approaches } \frac{R}{1.45} = 0.69 \, R.$$

This suffices for a graphical construction. Moreover, because the quotient (semi-chord)/R equals the sine of the opposite angle, our result implies that the sine of the angle of refraction is 0.69. Either a hand calculator or a table of sines informs us that the maximum angle is 44 degrees.

Now we study the transition plastic-to-air. We send in light along the semi-circle's radius so that the beam goes straight through from air to plastic, heading diagonally upward and to the left. We start at a small angle relative to the normal, just reversing the path shown in the lower half of figure 1.6(a). The angle there is now our angle of incidence. Observation 3, on reversibility, now comes into play. It tells us that, when the beam refracts at the plastic-to-air transition, the beam will emerge into air along the line already drawn in the figure. We get bending away from the normal, and all is well.

As we increase the angle at which we start the light beam, the refracted beam above the plastic becomes more nearly horizontal. What happens when we get to the angle of 44 degrees in figure 1.6(b) and then exceed that angle? Where does the light go?

There is no disaster, no sudden change. Nature is too subtle for that. As we increase the starting angle, more and more of the light is reflected, less and less is refracted. The amount refracted has decreased to zero when the starting angle reaches 44 degrees. Beyond that angle, there is no more refraction, just reflection. We have *total reflection*. Figure 1.7 illustrates the change in behavior.

The angle of incidence such that the angle of refraction equals 90 degrees is called the *critical angle*. For our plastic in air, the critical angle is 44 degrees.

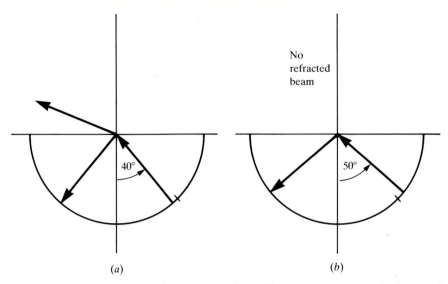

(a)   (b)

Figure 1.7 The disappearance of the refracted beam. The light is incident on the horizontal plastic-to-air interface from within the plastic. Each sketch shows the incident beam, the reflected beam, and – when it exists – the refracted beam. (a) When the angle of incidence equals 40 degrees, both reflected and refracted beams arise. (b) When the angle of incidence equals 50 degrees, only a reflected beam arises.

With the aid of this definition, we can codify the results as

Observation 7. When the angle of incidence exceeds the critical angle, the light beam is totally reflected.

In figure 1.7, the tick mark along the plastic's curved surface shows where one would start to draw an incident beam at 44 degrees, the critical angle. The tick mark divides the quarter-circle in the lower right into two regions: (1) a region where the angle of incidence is less than the critical angle and so both a reflected and a refracted beam arise, and (2) a region where the angle of incidence exceeds the critical angle and only a reflected beam arises.

Light pipes

Sending light in a straight line is easy. Getting it around corners – without losing intensity – is a different story. Even when a light beam reflects from a good silvered mirror, not all of the incident light bounces off; some is absorbed by the metal, as much as 20 per cent. A sequence of reflections, in which 20 per cent is lost at each bounce, will quickly reduce a powerful beam to a faint trace of its former self.

Total reflection solves the problem. Figure 1.8 shows a solid cylinder of glass bent into a gentle curve. A ray of light moving up the glass hits the

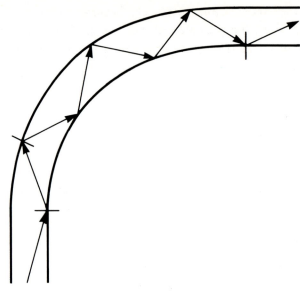

Figure 1.8 A light pipe. The three short lines perpendicular to the glass-to-air boundary are the normals to the surface at their respective locations. Despite what the name might suggest, the light pipe is solid glass, not hollow.

glass-to-air boundary at an angle of incidence greater than the critical angle. Hence the ray is totally reflected, and it proceeds on to its next encounter with the boundary. Provided the curve is gentle, the angle of incidence again exceeds the critical angle, and so the ray is again totally reflected. And on and on. Because there is no refraction of light out of the glass, there is no loss of intensity. The glass just "pipes" the light up and around the curve, much as a garden hose pipes water to the tomatoes.

In commercial use, light pipes are drawn out as very long, fine fibers of glass. The diameter is about 0.1 millimeters and the field of applications itself is called *fiber optics*. For example, telephone conversations between New York City and Chicago are routinely converted into digital information – a kind of Morse code – and that digital information is sent as tiny light pulses through light pipes connecting the two cities. The fiber, carefully encased, is laid underground where neither thunderstorms nor falling branches can interrupt the communication.

Intersection

Our next observation can be made in many contexts. You may have seen several searchlights sweeping the sky, advertising a carnival; their beams cross like sabres in the sky. Surely, on a foggy night, you have seen the headlights of two cars criss-crossing at an intersection. Or you can take two flashlights and let their beams intersect, forming the letter X. In none of these situations does the presence of one beam affect the behavior of the other, at least so far as the eye can perceive. Rather, we have

Observation 8. Light beams can intersect without perceptible effect.

1.2 The speed of light in vacuum

Thus far we have examined beams of light as they exist in space. We have not addressed the question of how light moves, in particular, how fast light travels. Direct observation is not easy, and not until the seventeenth century was the speed of light successfully determined.

When Newton composed his *Opticks*, which he published in 1704, he addressed the question of speed and wrote as follows:

> Light is propagated from luminous Bodies in time, and spends about seven or eight Minutes of an Hour in passing from the Sun to the Earth.
>
> This was observed first by Roemer, and then by others, by means of the Eclipses of the Satellites of Jupiter. For these Eclipses, when the Earth is between the Sun and Jupiter, happen about seven or eight Minutes sooner than they ought to do by the Tables, and when the Earth is beyond the Sun they happen seven or eight Minutes later than they ought to do; the reason being, that the Light of the Satellites has farther to go in the latter case than in the former by the Diameter of the Earth's Orbit.

Newton is referring to Olaus Roemer's calculations of 1676. Figure 1.9 sketches the scene.

If we could be on the planet Jupiter and travel with it, we would see Io, one of Jupiter's moons, circle Jupiter quite regularly. The time for a full circle is 42 hours and 28 minutes.

When Jupiter lies between Io and the sun, Io is eclisped; it no longer receives the sun's rays and no longer shines with reflected light. We would observe a periodic sequence of eclipses: Io entering Jupiter's shadow and then emerging.

The astronomers of the seventeenth century had studied Io ever since Galileo discovered the moons of Jupiter in 1610. They noted the eclipses and sought to predict their recurrence. But the eclipses – as observed on earth – did not recur perfectly periodically. If one used observations made when the earth was nearest to Jupiter to predict when eclipses will termin-

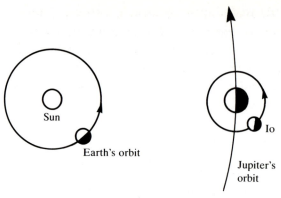

Figure 1.9 The planetary scene for measuring the speed of light. The drawing is *not* to scale. Jupiter takes so long to circle the sun – about 12 years – that its motion in 1 year plays no significant role in our analysis.

ate when the earth is farthest from Jupiter (about half a year later), then the eclipses actually seem to terminate later than calculated. The delay, reasoned Roemer, is due to the travel time for light to cross the earth's orbit.

Now we stand ready to determine the speed of light. Recall that speed means the rate at which something moves: distance traveled divided by elapsed time. Newton gives the time from sun to earth as 7 or 8 minutes. Let us average and take 15 minutes as the time for light to cross the diameter of the earth's orbit. In seconds, that interval would be 15 minutes times 60 seconds per minute, yielding 900 seconds.

Already in Newton's day, the diameter of the earth's orbit was known to be about 3×10^{11} meters.

Thus we may write

$$\text{speed of light in vacuum} = \frac{\text{diameter of earth's orbit}}{\text{apparent delay in eclipse termination}}$$

$$= \frac{3 \times 10^{11} \text{ meters}}{900 \text{ seconds}}$$

$$= \frac{30 \times 10^{10}}{9 \times 10^{2}} = 3 \times 10^{8} \text{ meters / second.}$$

To one significant figure, we find the speed of light in vacuum to be 3×10^{8} meters per second.

At this juncture, five points are worth noting. First, what we have here is the speed of light *in vacuum*. In the space between the earth and the sun, there is very little material: some stray hydrogen, a little dust, some cometary debris, and not much else. The region is much more nearly a perfect vacuum than any "vacuum" yet produced on earth, whether between the glass walls of a hot drink flask or in a laboratory at the leading edge of technology. What the speed of light may be in water or glass is – for us – an open question. The value 3×10^8 meters/second refers to the speed in vacuum.

Second, when we study the theory of relativity, we will need a convenient symbol for the speed of light in vacuum. The letter c is now commonly used:

$$c = \text{speed of light in vacuum}$$
$$= 3 \times 10^8 \text{ meters/second.}$$

Why "c"? Probably the letter was taken from the Latin *celeritas*, swiftness, a root that gives us the word *celerity*.

Third, light is indeed swift. Because 3×10^8 equals 300×10^6, light travels 300 million meters per second. With an appropriate set of mirrors, we could – in 1 second – send light around the earth seven times.

Fourth, when Roemer interpreted the time delay in Io's eclipses as being the transit time for light, he adduced evidence that light travels at a *finite* speed. Light does not get from point A to point B instantaneously. The finiteness of the speed is what Newton had in mind when he wrote that "Light is propagated from luminous Bodies in time." It takes time for light to get from A to B.

Fifth, we assessed the speed of light using information available to Newton in 1704. Determinations of c made in the nineteenth and twentieth centuries provided accuracy to many decimal places, but we will rarely need the additional accuracy. Like our value from 1704, the modern values round off to 3×10^8 meters/second, and that simple number will suffice almost always.

1.3 White light and colors

We return now to observations that can be made in a classroom and ask, what happens if we send light through a wedge-shaped piece of glass, that is, through a triangular prism?

Figure 1.10 suggests the answer. Like water, glass has an index of refraction greater than 1. So when the beam first strikes the glass, the beam will be bent toward the normal. After traversing the glass, the beam will make a transition from glass to air, the reverse of the first transition. Observation 3,

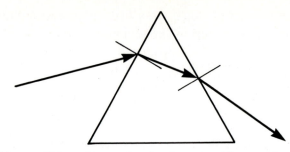

Figure 1.10 A beam of light striking a triangular glass prism. The second and third arrows suggest what will happen — according to our understanding of refraction thus far.

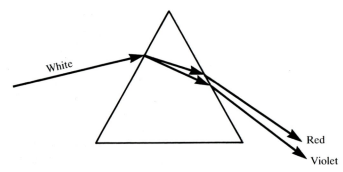

Figure 1.11 When a glass prism spreads white light out into a spectrum, the red and violet rays take separate paths. (For clarity's sake, the divergence of the two paths has been exaggerated.)

on reversibility, implies that now the beam will be bent away from the normal.

An opaque slide with a long, narrow slit cut in it, when placed in a 35-millimeter slide projector, produces a narrow wafer of white light. When that wafer-like beam strikes the glass prism, the angles of refraction are pretty much as we expected – but the beam spreads out into a lovely spectrum of colors, red on the top, then the sequence orange, yellow, green, and blue, and finally violet at the bottom.

Figure 1.11 shows the paths for rays of the extreme colors, red and violet. The white light with which we started has been spread out into all the colors of the rainbow. With some boldness of inference, we summarize this experiment as

> Observation 9. White light consists of a mixture of all the colors of the rainbow.

We are not yet done; figure 1.12 shows our next experiment. In the region of space where the spectrum appears, we place a large piece of

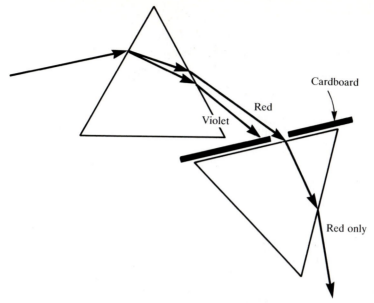

Figure 1.12 Because of the cardboard baffle, the second prism receives only light of a single color – and refracts it without any change in color.

cardboard with a narrow slit in it. The paper blocks all the light except for a new narrow wafer, all of one color (or very nearly so). In the figure, only red light is allowed to pass. In the path of that light, we place a second prism. What happens now?

The red light is refracted at both surfaces and emerges as red. There is no change in color. We shift the cardboard sideways and try orange, yellow, green, blue, and violet. Each distinct color is refracted twice and emerges as the same color. Passage through the glass prism does not change the color of the light that falls on it. That being so, we may be confident that the action of the first prism is merely to sort out a mixture of already-existing colored lights.

As figure 1.11 already indicated, the violet component of white light is bent more by a glass prism than is the red component. That a prism produces a spectrum does not force us to discard Snell's law, as we have understood it so far, codified in Observations 5 and 5'. Rather, we must refine it. We need to make

> Observation 10. Snell's law still describes refraction correctly, but the amount of refraction for a given angle of incidence varies with the color of the light. An equivalent statement is this: the index of refraction is different for different colors of light. For example, the index has one numerical value for red light and a different value for violet light.

Table 1.2. *How the index of refraction varies with color. The water is at room temperature; the glass is ordinary window glass.*

	Color					
	Red	Orange	Yellow	Green	Blue	Violet
Water	1.331	1.333	1.333	1.335	1.338	1.341
Glass	1.514	1.517	1.518	1.520	1.524	1.529

In particular, violet light is bent more than red light by a glass prism.

Table 1.2 provides numerical information.

Why had we not noticed this effect before? Because the amount of refraction varies only slightly with color. For example, in the case of water, the index of refraction for red light is 1.331, while for violet it is 1.341. When a broad beam of light leaves a tank of water, as in part (*a*) of figure 1.3, the red and violet components do emerge at slightly different angles. Close to the tank, however, those two colored beams overlap each other and largely coincide with the beams of all the intermediate colors. The resulting mixture is still perceived as white light. Only far away will the divergence in directions have spread the beams significantly apart. Indeed, when a cylindrical beam of light refracts out of water, you may have noticed a reddish cast to one crescent of an out-of-round "white" spot on a wall and a bluish cast to the opposite crescent. The content of Observation 10 is beginning to manifest itself.

1.4 Sidedness

If you own a pair of clip-on sunglasses, you may have tried the following experiment. You hold the sunglasses about 20 centimeters in front of your face and look at a bright, predominantly-horizontal scene: the beach, a concrete superhighway, the rippled surface of a lake, or the out-run of a ski slope. When you look through just one lens and rotate it, the brightness of the scene changes. A rotation of 180 degrees returns the brightness to its original value.

Before we consider another experiment, let me tell you how the material in a typical pair of clip-on sunglasses is produced.

The manufacturer begins with polyvinyl alcohol, a very long molecule made of identical small units. (The unit, vinyl alcohol, has the chemical formula CH_2-CHOH, but that information is just an aside.) Many of these long molecules, mixed together, form a sheet of clear plastic. The molecules themselves are in helter-skelter disarray, like a serving of spaghetti spread out over a large plate. The manufacturer aligns the long molecules (at least partially) by stretching the plastic sheet while it is hot. What the result looks like is sketched in figure 1.13. The stretched sheet of polyvinyl alcohol is cemented to a stiff sheet of plastic; this prevents unstretching and loss of alignment.

Figure 1.13 The manufacture of Polaroid material. Stretching has partially aligned the long molecules of polyvinyl alcohol.

Next, the composite sheet is dipped into a solution rich in iodine. The iodine diffuses into the polyvinyl alcohol and combines with it, forming a string of iodine atoms along each molecular chain.

The chemistry in this process, however vital in practice, is not the essential feature for us. Rather, the stretching and consequent alignment are central. No longer are all directions in the sheet equivalent. A special direction has been embedded in the plastic, the direction of the stretching and alignment.

Now we are ready for a fruitful experiment. We take two large sheets of the material and hold them up to the light, one over the other and with their "stretch directions" parallel to each other. Part (*a*) of figure 1.14 shows this. The light comes through quite well. Holding the more distant sheet fixed, we rotate the closer one; less and less light is transmitted to our eyes. As we approach 90 degrees, the scene goes black. Then, as we pass 90 degrees and head toward 180 degrees, brightness returns. Except for two orientations, the closer sheet absorbs some of the light, and the fraction that it absorbs varies – smoothly but strongly – with changes in orientation.

This is the observation in its primitive form. Let us think about it a moment, focussing our attention on the light as it speeds *between* the two sheets. For simplicity, make it a cylindrical beam of light from a projector behind the more distant sheet. When we rotate the sheet closer to us, the intensity reaching us changes. Unless the beam of light has some kind of

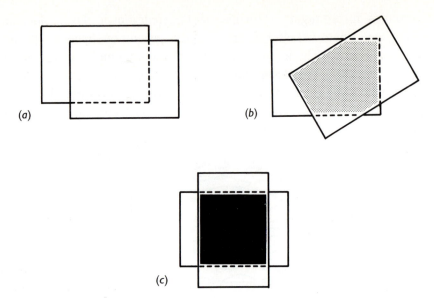

Figure 1.14 Light reveals its "sidedness" when we rotate one sheet of Polaroid in front of another. (*a*) Stretch directions parallel; (*b*) oblique; (*c*) at right angles to each other.

intrinsic "sidedness," our rotating the closer sheet could have no effect on the intensity reaching us. After all, the closer sheet is always there; we are merely changing the orientation in space of the "stretch direction."

Nothing in our experiment tells us what the nature of this sidedness is, but we are entitled to make

Observation 11. Light has some kind of "sidedness."

Later, in chapter 5, we will be able to clear up this mystery.

1.5 Variations in intensity

Next, we revisit an aspect of light that we met earlier. Figure 1.15 shows a wafer of white light incident on a prism. Some of the light is reflected when it first encounters the glass. The rest of the light is refracted into the glass, and some of that light is then refracted out the far side, producing a spectrum. Why is some light reflected and some refracted? What determines the relative amounts, reflected and refracted? When we discovered total reflection (codified in Observation 7), we saw clearly that the amount reflected changes with the angle of incidence.

Two good questions, but we cannot answer them yet. Rather, another experiment dramatizes the unresolved issues.

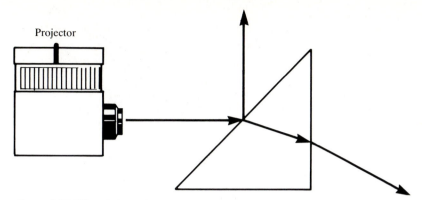

Figure 1.15 When light strikes a glass prism, we get both refraction and reflection.

The mineral mica can be split into extremely thin sheets, each quite clear and transparent. Newton knew the material as "Muscovy glass" because large sheets were used for the window panes of houses in old Russia. Imagine a mica sheet the size of your two hands, held together with the palms outward. The thickness of the sheet, however, is only 5/100 of a millimeter, about the thickness of a sheet of fine paper.

For a light source, we need something that gives us only a single color. Newton used a prism to spread white light into a spectrum; then he discarded the colors that he did not want, keeping only one, but the final intensity is woefully low. We will do much better with a modern sodium vapor lamp. The element sodium is a metallic solid at room temperature. When heated, it melts, and a vapor of sodium atoms is formed, too. The hot vapor produces a soft yet intense glow of light, yellow with an orange cast. (Some modern street lights – those that produce an orangish light – work in much the same fashion.)

When we hold the mica sheet 10 centimeters from the sodium lamp, some light is transmitted through the sheet, and some is reflected. Figure 1.16 shows some paths for the light. The pattern of bright and dark on a wall (distant by several meters) is remarkable. A bull's eye pattern consisting of a dark central disk surrounded by concentric bright rings on a dark background forms in one kind of light, either transmitted light or reflected. Figure 1.17 shows the former situation. The complementary pattern, a bright central disk followed by concentric dark rings on a bright background, forms in the light that travels in the opposite direction.

We can enjoy the beauty of the patterns without troubling ourselves (at this time) about an explanation. Indeed, we close this section with

> Observation 12. When light reflects and refracts, the total intensity is split in some fashion. In particular, when light of a single color impinges on a

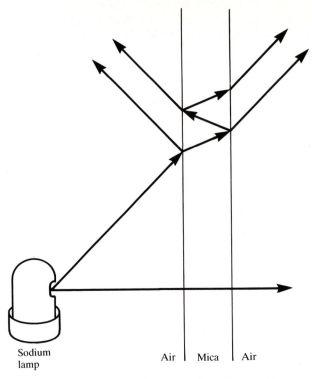

Sodium
lamp
Air | Mica | Air

Figure 1.16 Some paths for light striking the mica sheet.

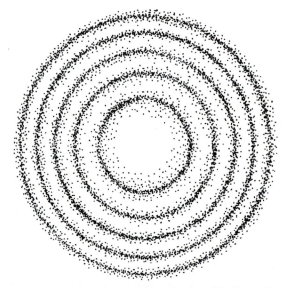

Figure 1.17 The pattern of bright and dark on a distant wall when yellow light impinges on a thin sheet of mica. A dense set of dots represents brightness; the absence of stippling represents darkness. The pattern was formed by transmitted light in an otherwise dark room.

thin sheet of mica, the patterns of reflected and transmitted intensity are daunting in their complexity but lovely in their appearance.

1.6 The rainbow

Of all the optical phenomena in everyday life, the rainbow is the loveliest. And the observations in this chapter suffice for understanding how the rainbow arises. We do not need a theory of what light *is* in order to understand how sunlight and raindrops produce a rainbow.

Figure 1.18 shows someone – her name is Alice – looking at a rainbow. Alice is looking into a region where raindrops are falling, and the sun is at her back. The rainbow makes a glorious arc in the sky. Reflection of sunlight by the raindrops is certainly an essential element of an explanation, but refraction – it turns out – plays a role, too.

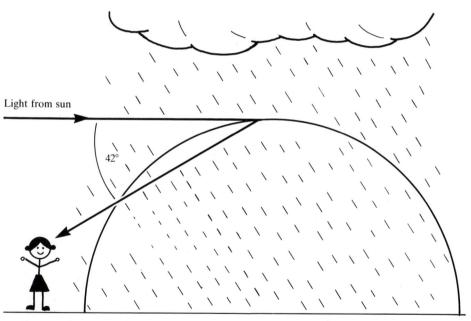

Figure 1.18 Alice looking at the red arc in the (primary) rainbow. The angle between her line of sight and the direction of the incident sunlight is 42 degrees.

Figure 1.19 shows the crucial path of light. A ray from the sun strikes the spherical raindrop, and some light is refracted into the water. (Here we may ignore the portion that is reflected by the drop's surface.) Next, the ray proceeds to the far side of the drop and is reflected there. (Now we may ignore the portion that is refracted.) Finally, the ray strikes the underside of

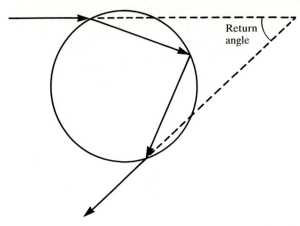

Figure 1.19 The crucial path of a light ray for producing the (primary) rainbow. The circle represents the cross-section of a spherical raindrop. For the light ray, the sequence is refraction, reflection, and refraction. The angle (less than 90 degrees) between the incident direction and the emergent direction we will call the *return angle*.

the drop and is refracted out into the air. (Whatever light is reflected this time is of no interest right now.)

In precisely which direction does the emergent ray travel? Two rules – Snell's law for refraction and the equality of the angles of incidence and reflection – suffice for answering that question, once the initial point of contact between the raindrop and the ray from the sun has been specified. Also, because the index of refraction depends on color, we must specify the color of the light. For a start, let us take red. Working out the complete path for many rays – that is, for many different initial points of contact – reveals a surprising geometric property: the return angle for red light never exceeds 42.5 degrees, and most rays have a return angle near 42 degrees. Figure 1.20 illustrates some of this. In short, the reflection – which is preceded and followed by a refraction – concentrates the red light into return directions with a return angle near 42 degrees.

When Alice looks at raindrops for which the angle between her line of sight and the direction of the incident sunlight is 42 degrees, red light is strongly reflected back to her. When she looks at raindrops for which the angle is different by even 1 degree, almost no red light is reflected in her direction. Thus Alice sees a red arc in the sky. The arc is formed by all drops that lie on the surface of a cone which has its apex at Alice's eyes, as illustrated in figure 1.21. The axis of the cone is parallel to the direction of the incident sunlight.

When we repeat the computations for violet light, we need to use a different value for the index of refraction, the value appropriate to violet,

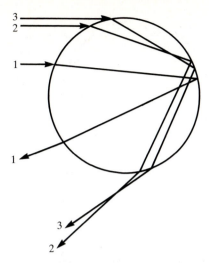

Figure 1.20 The paths of some representative rays of red light. Ray 1 strikes the raindrop near its center and therefore comes almost straight back; it has a small return angle. Although ray 3 strikes the raindrop farther off-center than does ray 2, it has a smaller return angle than does ray 2. Indeed, if one were to draw other rays, none would have a return angle greater than does ray 2, whose return angle is about 42 degrees.

For us to see the concentration of emergent rays near ray 2, many rays would have to be drawn, and the diagram would be hopelessly cluttered. So that aspect is not illustrated.

A Florence flask filled with water provides a fine large-scale model of a raindrop. A diameter of about 10 centimeters works well. By shining a red laser beam at the flask and varying the point of initial contact, one can see that the return angle has a maximum around 42 degrees. By illuminating the entire flask with a broad beam of (white) light from a projector, one can see that the emergent light is concentrated near the maximum return angle.

rather than to red. The same general behavior emerges – the reflected violet light is concentrated – but now the return angle for the concentrated light is about 40 degrees. Figure 1.22 shows the consequences for Alice.

Succinctly, because the index of refraction varies with color, so does the return angle at which light is concentrated. Thus different colors come to Alice from different sets of raindrops and produce a colored rainbow.

This analysis suffices for one rainbow, but occasionally you see two rainbows, one higher in the sky than the other. The lower, brighter rainbow arises – as just described – from one reflection, preceded and followed by refraction, and is called the *primary* bow. The higher, dimmer rainbow arises from two reflections inside the drop, again preceded and followed by refraction, and is called the *secondary* bow. Next time you see both rainbows, look to see whether the sequence of colors is the same in them.

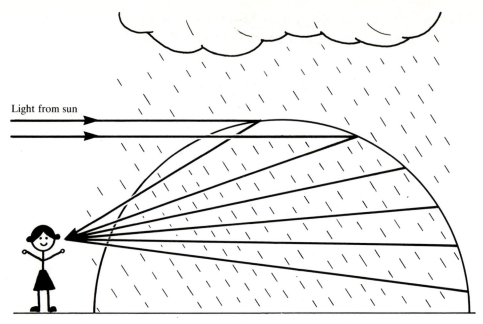

Light from sun

Figure 1.21 A sketch suggesting the cone. The lines that reach Alice's eyes lie on the cone's surface. A drop anywhere along any one of these lines will reflect red light toward Alice.

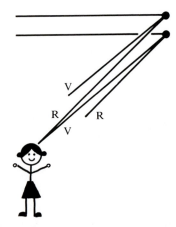

Figure 1.22 Both raindrops reflect red and violet light from the sun. (The return angles are about 42 degrees for red (denoted by R) and 40 degrees for violet (denoted by V).) Nonetheless, only the red light from the higher drop and the violet light from the lower drop reach Alice's eyes. Thus Alice sees a violet arc inside a larger red arc.

The intermediate colors in the spectrum of white light – orange, yellow, green, and blue – have intermediate values for the index of refraction, and so they appear in intermediate arcs between the red and violet arcs.

Figure 1.22 shows that the primary bow has red on the outside of the broad arc and violet on the inside. But what about the secondary bow?

Additional resources

Vasco Ronchi's *The Nature of Light: An Historical Survey* (Harvard University Press, Cambridge, MA, 1970) provides a charming history of light, all the way from the Greeks to today.

More about the rainbow can be found in Robert Greenler's *Rainbows, Haloes, and Glories* (Cambridge University Press, New York, 1980), a vivid and engaging exploration of the topics in the title.

Square sheets of mica, 15 centimeters along an edge, can be purchased (under the name "Muscovite" and at a modest price) from Ward's Natural Science Establishment, P.O. Box 92912, 5100 West Henrietta Road, Rochester, New York 14692–9012. Ask for a parallel sheet.

Questions

1. If you can see the eyes of someone in a complicated system of mirrors, is it possible for that person to see your eyes? Yes or no, and then explain your reasoning.

2. Figure 1.23 shows the path of light traveling from air into glass. Is the glass on the right or on the left in the drawing? Explain your reasoning (briefly).

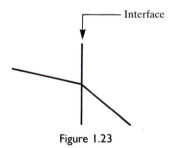

Interface

Figure 1.23

3. Here are some review questions about working with powers of ten.

(*a*) Write 2000 as 2 times $10^{\text{some exponent}}$

(*b*) $3 \times 10^3 + 2 \times 10^3 =$

(*c*) $3 \times 10^3 + 2 \times 10^2 =$

(*d*) $\dfrac{6 \times 10^5}{2 \times 10^2} =$

(*e*) Which is larger, 2×10^3 or 5×10^{-7}?

(*f*) $\dfrac{6\times10^5}{2\times10^{-4}}=$

(*g*) $(10^3)^4=10^{\text{What number goes here?}}$

4. Suppose we shine a beam of light (as in figure 1.24) straight at the center of an air bubble that is under water. (If you worry about the bubble's floating way, think of a transparent balloon filled with air and tied in place.) Does the light beam actually go through the center of the bubble? Sketch the actual path to the distant side of the bubble (on the assumption that the light does at least enter the bubble.) Be careful and consistent.

A point worth noting. The "normal" is a line perpendicular to a surface at the transition point. Thus the "normal" to a spherical surface is along that radius of the sphere which passes through the transition point.

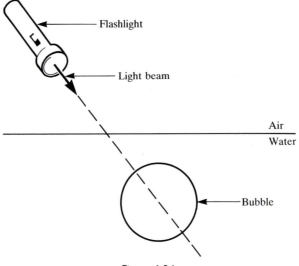

Figure 1.24

5. For the transition air-to-glass, the index of refraction is 1.5. (That is a good typical figure; the precise value varies with the composition of the glass and the color of the light.) Reproduce the sketch in figure 1.25 and add the reflected and refracted rays. A *careful* graphical treatment is all that is asked for, but you will need to do some calculating (with the index 1.5) before you can draw the refracted ray faithfully.

6. In 1987, astronomers were treated to the first supernova visible to the naked eye in more than 300 years. When certain stars become old, their evolution carries them through an explosive phase. Suddenly the star

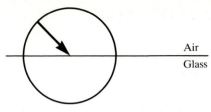

Figure 1.25

becomes about a billion times brighter than normal, and the appearance of this anomalously bright star in the heavens is called a supernova, a "new" star that is superbright. Stars are very distant, and that is particularly true of all supernovas within human memory. When the bright light from a supernova reaches us, it has been traveling for both a long distance and a long time. (For the 1987 supernova, the time was about 160 000 years.) Yet we see the explosion as a bright *white* light, not as a series of different colors arriving at different times.

> (a) What inference can you draw about the speeds with which different colors of light travel through interstellar space (which is virtually a vacuum)?

> (b) A triangular glass prism splits white light into a spectrum, different colors being bent through different angles. Suggest one or more ways in which a particle theory of light can explain this phenomenon and yet be consistent with a single speed of light for all colors in vacuum. (Part (b) is most appropriate after you have studied the first two sections of chapter 2.)

7. Imagine that two mirrors are put together at a right angle (as in both parts of figure 1.26). You are looking down on the mirrors, as though on books opened 90 degrees.

> (a) In sketch A, a beam of light approaches the right-angled mirror parallel to the center line (or symmetry line). What is the direction of the beam when it emerges? Provide a sketch of the full path.

> (b) In sketch B, the beam comes in at an angle (relative to the center line). Now what is the beam's direction when it emerges? Try a *careful* drawing as part of your response.

You have just worked out in two dimensions the principle that is employed (in three dimensions) in the laser reflectors that were placed on the moon in 1969, when astronauts first visited that extra-terrestial sphere.

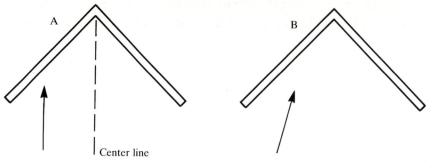

Center line

Figure 1.26

8. Near the end of section 1.1, we discovered total reflection in the context of light passing from plastic into air. Is total reflection possible for light striking a water–air interface from the water side? From the air side? How would you generalize from these specific pairs of materials?

9. Observation 2(*b*) noted that the incident ray, the reflected ray, and the normal lie in the same plane. The text never addressed the analogous issue for the refracted ray. To be sure, Snell's law tells us the angle that the refracted ray makes with the normal, but how is the direction of the refracted ray related to the plane defined by the incident ray and the normal? Does the refracted ray lie in that plane (as does the reflected ray)? Experiment is the only sure way to tell, but you can make some shrewd guesses in advance.

(*a*) Suppose the material in which the refracted ray travels is a gas or liquid, in which the molecules have no definite spatial pattern or arrangement but rather are in helter-skelter disarray. Could nature choose any direction for the refracted ray other than in the plane defined by the incident ray and the normal?

(*b*) Consider now a crystalline solid, a material in which the atoms or molecules are arranged in a regular pattern, like bricks in a wall or offices in a skyscraper. Ask yourself the same question as in part (*a*).

(*c*) Repeat part (*b*) for an amorphous solid, a substance like glass, in which there is no long-range regular pattern for the molecules.

10. A narrow beam of light impinges (from air) on one surface of a triangular glass prism. What information would you need in order to predict the path of the beam through the prism and out again? Use a sketch as an aid in presenting your response. For definiteness, specify that the triangle has sides of three different lengths and that the beam strikes the shortest side.

11. A beam of light travels from air into water.

(*a*) If the original angle of incidence is 10 degrees and you double the angle of incidence, will you double the angle of reflection? Will you double the angle of refraction? (Exactly? To good approximation?)

(*b*) Repeat part (*a*) but with an original angle of incidence of 40 degrees.

12. Refer to figure 1.1 and its beam of light traveling from air into water. The angle of incidence is 30 degrees. You are now given two additional facts: (1) the water is 20 centimeters deep; (2) the refracted beam strikes the bottom 8.1 centimeters beyond the point that is directly below the location where the light enters the water.

(*a*) What is the sine of the angle of refraction?

(*b*) What is the angle itself?

(*c*) What value do you compute for the index of refraction of water?

13. It is dark outside, and you are in a brightly-lit room. When you look out the window, you see the reflection of objects located in the room, yourself included. Although there is only a single pane of glass, often you see each object doubled. Why is that?

14. Suppose you shine a narrow beam of light at one of the vertical sides of a rectangular aquarium. The light travels through air, then through glass, and finally through water.

(*a*) If the beam in air travels perpendicular to the glass, how does the beam travel in the water?

(*b*) Now specify that the beam in air travels at an angle of incidence of 20 degrees (relative to the glass). Sketch the path through air, glass, and water. You need not compute exact angles, but be faithful to the qualitative behavior. Does the beam in the water travel parallel to the beam in air?

15. Beams of monochromatic red light enter two glass prisms, first arranged base to base, as in part (*a*) of figure 1.27 and then tip to tip, as in

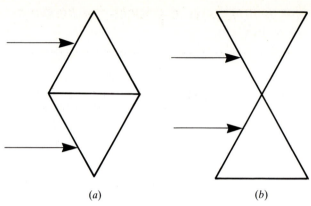

(a)   (b)

Figure I.27

part (b). For each configuration, trace qualitatively the paths of the light beams through the prisms and out the other side. If you can see a connection between these configurations of prisms and any lenses with which you are familiar (either convex or concave), describe the link.

2 Newton's particle theory

Even the Rays of Light seem to be hard bodies; for otherwise they would not retain different Properties in their different Sides.

Isaac Newton,
Query 31, *Opticks*

2.1 Theory building

By now we have a substantial number of observations; we know quite a bit about how light behaves. Let us try our hand at building a theory of light: what light is, and why it behaves as it does.

Our first observation – that light goes in straight lines from a luminous source – suggests that we need the notion of something that moves through space. The simplest thing is a particle, a "little baseball." Perhaps the luminous source emits little light particles, a stream of them. Figure 2.1 illustrates this notion.

Isaac Newton developed the same idea. He worked on light – experimentally and theoretically – most of his long life. At the start of chapter 1, we noted that Newton's first paper (published in 1672) was on optics, and his interest in the subject goes back at least to his student days at the University of Cambridge, England, perhaps to the year 1663.

Newton could not abide criticism, and so he went to great lengths to avoid even the possibility of it. Because Robert Hooke and he readily came to acrimonious disagreement, Newton withheld the publication of his book on optics until after Hooke's death. Finally, in 1704, Newton published his treatise *Opticks*. It was written in English, rather than Latin, and hence was more accessible to readers (within England) than the standard Latin of the day (though, to be sure, every educated person knew Latin). Descriptions of experiments filled its pages, and Newton took pains to explain the apparatus carefully enough that others could repeat his experiments. Intriguing conjectures closed the book; they laid out Newton's deepest thoughts about the nature of light and the cause of gravity. The book was easy to read – especially in comparison with Newton's earlier and even greater work, *Philosophiae Naturalis Principia Mathematica*, the Mathe-

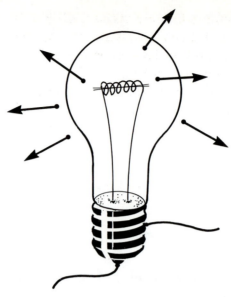

Figure 2.1 Particles of light in Newton's conception. The hot filament of a 100-watt bulb emits little particles of light, particles akin to little baseballs.

matical Principles of Natural Philosophy. For the *Opticks*, widespread popularity was instantaneous.

Newton revised the book three times; the fourth edition appeared posthumously in 1730. The number of conjectures – called "queries" by Newton – grew with time, from sixteen in the first edition to thirty-one in the fourth.

These historical remarks are interjected because, in this chapter, we construct a theory of light that is actually Newton's theory, and so quotations from his *Opticks* appear frequently. Indeed, in his Query 29, Newton asked, "Are not the Rays of Light very small Bodies emitted from shining Substances?"

Let us see whether we can build a theory of light based on the particle notion. Specifically, can we account for the behavior of light by supposing:

(*a*) that light is made up of little particles; and

(*b*) that those particles obey the same laws of physics as do other objects, for example, baseballs, whatever those laws may be in detail?

We can state the question in another (but equivalent) way: can we build a *particle theory of light* based on the laws of motion that Newton developed for large-scale objects, things like cannon balls and planets?

Why should we specify that the particles of light be little? Because our Observation 8 tells us that light beams intersect without interaction (at least at our level of perception). If the light particles are very small, then, when beams of light intersect, the probability of collison is small, imperceptibly small. In contrast, when many cars cross a busy intersection where the traffic light is out of order, fenders are likely to be bent. Cars are not tiny objects relative to the width of the street or the distance between cars in city traffic.

What about Observation 2, on reflection? Suppose we throw a small superball (perhaps 4 centimeters in diameter) at a table top with an "angle of incidence" of 30 degrees. How does the ball bounce? The "angle of reflection" is roughly 30 degrees, and the motion is confined to a single vertical plane. Yes, it is possible to imagine "little balls" that bounce just as light reflects.

Good, but all balls that we can pick up and toss are pulled on by gravity and go in curved paths on the earth. Doesn't this put us in conflict with the straight-line motion of light (in a single substance)?

Maybe – but maybe not. Take a can of tennis balls and throw the balls horizontally, the first ball quite slowly, the second much faster, and the third as hard as you can. The slow ball traces a prominently curved path to the ground; the very fast ball shows little curvature.

Perhaps light travels so fast that the curvature, though existing, is ordinarily imperceptible.

In section 1.2, we learned that Newton had a good value for the speed of light: 3×10^8 meters/second, when expressed in modern units. That speed is so remarkably high that we would expect extremely little curvature as a light particle traversed the length of a room. Our observations about straight-line motion do not pose a difficulty for a particle theory of light.

2.2 The particle theory of refraction

Now we consider refraction. What could make a light particle change its path as it goes from air to water? A pull in the vicinity of the surface could.

Figure 2.2 sketches the context. When a light particle is thoroughly surrounded by one material or the other, there ought to be no net push or pull. The symmetry of the situation – as much material above as below or to the right as to the left – assures us of that.

In physics, the term *force* means a push or a pull, and so we can say that when a light particle is far from the surface, no net force acts on it.

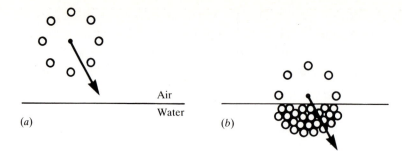

Figure 2.2 How a force may arise on a light particle near the surface of water. Molecules of air and water are represented by the small open circles. The arrows indicate the light particle's motion. (*a*) No net force. (*b*) More stuff below than above and hence a net force.

Near the surface, however, an imbalance of pulls is possible, and such an imbalance would deflect a particle from its original path.

If water pulls more on a light particle than does air, we can account for Observation 4, that the angle of refraction is *less* than the angle of incidence on the transition air-to-water, that is, that the light beam is bent toward the normal.

An experiment with a steel ball will illustrate the bending well. Roll the ball along a smooth surface that leads to a ramp, as sketched in figure 2.3. A drop of 5 centimeters over a distance of 15 centimeters will do nicely. While the ball is on the flat table, approaching the ramp, it moves with constant speed. While going down the ramp, however, the ball experiences a force (from the ramp) that is partly horizontal as well as partly vertical. The horizontal portion increases the ball's horizontal speed. Moreover, if the ball was headed toward the ramp in some diagonal direction, the horizontal portion of the force deflects the ball's trajectory: the ball's path bends. On the lower horizontal surface, the ball again moves with constant speed – but with a higher speed, and its course has changed.

Now we return to light. A net pull near the surface will bend the path of a light particle; specifically, it will bend the trajectory toward the normal. This provides a qualitative explanation of refraction.

But how about the quantitative behavior, the numerical detail?

Refraction quantitatively

We are dealing explicitly with motion in two (or three) dimensions, and so we need to specify direction just as carefully as speed. The technical term we need is *velocity*. To cite the velocity of an object is to specify both the

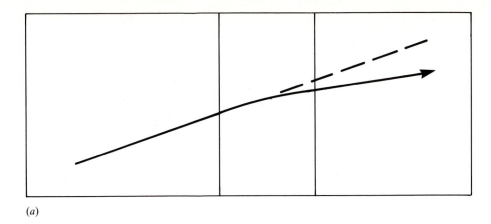

(a)

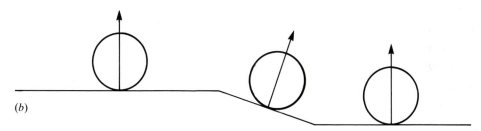

(b)

Figure 2.3 A steel ball rolling along. In (a), we see the table and ramp from the top; in (b), from the side. As the ball descends the ramp, part (a) shows how its trajectory bends. In part (b), the arrows indicate the force which the table and ramp (approximated as frictionless) exert on the ball.

direction of motion and the speed. In short, velocity answers the questions, "whither?" and "how fast?"

We can represent velocity by an arrow whose direction is the direction of motion and whose length is proportional to the speed. We denote the speed itself by the letter v (for the speed is the magnitude of the velocity). Figure 2.4 illustrates what we will mean by a *component* of velocity.

Now we can give our attention to refraction. Figure 2.5 shows us the context. The question for us is this one: given the incident ray, how should we draw the refracted ray? We can answer that question by determining what should happen to each component of the velocity.

Because the net force is perpendicular to the surface, the component of velocity parallel to the surface should not change.

For the same reason, the perpendicular component of velocity should increase.

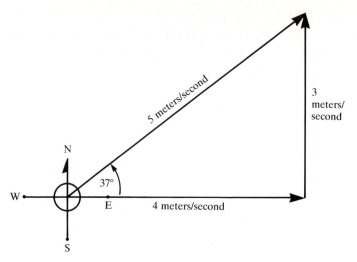

Figure 2.4 Velocity and its components. To specify the velocity of a runner, we can say the runner is moving 37 degrees north of east at a rate of 5 meters/second. That velocity will carry the runner 4 meters to the east in 1 second and 3 meters to the north (because 37 degrees gives a 3–4–5 right triangle). The amount of velocity in the eastward direction is thus 4 meters/second; we call it the eastward *component*. Similarly, the northward component is 3 meters/second. The actual motion at 37 degrees is equivalent to the runner's moving – simultaneously – eastward at 4 meters/second and northward at 3 meters/second.

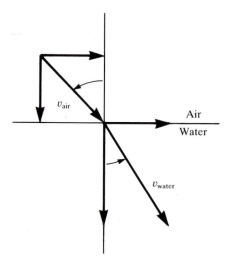

Figure 2.5 Refraction according to Newton's particle theory. Each arrow is drawn proportional to the corresponding velocity or component of velocity. Short arcs indicate the angles of incidence and refraction. (To forestall a misconception, let me mention here that gravity plays no role in refraction. Light is bent toward the normal if it enters a tank of water by passing upward through a thin transparent bottom or by traveling horizontally through transparent sides, just as it is bent toward the normal when entering from the top, as in this sketch.)

These statements are incorporated in the figure, and just by looking at the diagram, we can extract an implication. Because the arrow representing the speed in water is longer than the arrow for the speed in air, we infer a greater speed in water than in air:

$$v_{\text{water}} > v_{\text{air}}.$$

Observation 5′ was couched as a ratio of sines. What emerges here if we form the ratio? After consulting figure 2.5, we can write

$$\frac{\sin\,(\text{angle of incidence})}{\sin\,(\text{angle of refraction})} = \frac{\dfrac{\text{parallel component}}{v_{\text{air}}}}{\dfrac{\text{parallel component}}{v_{\text{water}}}}$$

$$= \frac{v_{\text{water}}}{v_{\text{air}}}.$$

Because the parallel component of velocity remains constant during the transition, we may cancel it in numerator and denominator. Then multiplication of both numerator and denominator by v_{water} permits a cancellation in the denominator, and the final form emerges. The ratio of sines depends only on the ratio of speeds.

Moreover, Newton was able to prove that the increase in speed (air to water) due to the pull would *not* depend on how the light particle started across the boundary. All he had to assume was that light acts like a tiny baseball and obeys the laws of motion which he had formulated for large-scale objects, things like pendulum clocks, cannon balls, and planets. Thus the ratio of speeds is a constant, independent of the angle of incidence, and Newton's particle theory neatly accounts for Snell's law.

Newton, too, was pleased. When he took up refraction in the *Opticks*, he said he would suppose "That Bodies refract Light by acting upon its Rays in Lines perpendicular to their Surfaces." He noted that he "must distinguish the Motion of every Ray into two Motions, the one perpendicular to the refracting Surface, the other parallel to it," just as we did. After these remarks come two pages of calculations: Newton showed that the ratio of sines is equal to a ratio of speeds and, in turn, that the ratio of speeds is the same for all angles of incidence. He concluded with the words,

> And this Demonstration being general, without determining [that is, without presupposing] what Light is, or by what kind of Force it is refracted, or assuming any thing farther than that the refracting Body acts upon the Rays in Lines perpendicular to its Surface; I

take it to be a very convincing Argument of the full truth of this
Proposition,

namely, that the empirical Snell's law provides an exact statement of how
light refracts.

2.3 Ramifications

With the basic mechanism of refraction thus understood, we can readily
make theoretical sense of several other observations.

Our Observation 6 noted that the index of refraction depends on the pair
of materials. Nothing is easier to explain.

Different materials exert different pulls, surely, and so the net pull
depends on the pair of materials, for example, air-to-glass versus air-to-
water. This change in net pull will alter both the amount of deflection (at a
given angle of incidence) and the change in speed. In short, the change in
net pull will alter the index of refraction.

Observations 9 and 10 concern colors. How can we account for the very
existence of different colors in a particle theory? And for different colors
being bent through different angles (by the same pair of materials)? Violet,
we noted, is bent more than red by a glass prism.

If you pause a few minutes to ponder these questions, various possibilities
will probably come to mind.

Here is what Newton had to say in his Query 29:

> Nothing more is requisite for producing all the variety of Colours,
> and degrees of Refrangibility [meaning "refraction"] than that the
> Rays of Light be Bodies of different Sizes, the least of which may
> take violet the weakest and darkest of the Colours, and be more
> easily diverted by refracting Surfaces from the right [meaning
> "straight line"] Course; and the rest as they are bigger and bigger,
> may make the stronger and more lucid colours, blue, green, yel-
> low, and red, and be more and more difficultly diverted.

Newton suggests light particles of *different sizes*. Particles of red light are
bigger and more massive than particles of violet light. Hence the particles of
red light would be deflected less by the same force acting near the surface.

Observation 11 indicated that light has some kind of sidedness. When we
rotate one Polaroid sheet in front of another, we vary the transmitted
intensity, from bright to dim to nothing and then back through the
sequence.

In our present theoretical frame of mind – little particles – we note that rotation of the Polaroid sheet which is closer to us could not change the transmitted intensity if the particles were spherical. For example, suppose you hold a bright yellow beach ball firmly in the palm of your left hand. You place your right hand over the ball, fingers outstretched. Then, holding the ball fixed with your left hand, you rotate your right hand 30 degrees. If you focus your attention on solely the ball and your right hand, then you do not notice any change in the orientation of your hand relative to the ball – precisely because the ball is a perfect sphere. There are no landmarks on it by which to register a rotation of your hand. The fingers of your right hand are like the "stretch direction" in a Polaroid sheet. If light particles were spherical, rotation of the sheet would have no perceptible effect.

Thus far, we conclude that light particles cannot be spherical. What did Newton have to say? Of course, Polaroid sheets were not available in 1704, but an equivalent phenomenon was known to Newton: the bizarre behavior of light when it passes through crystalline calcium carbonate, then called Iceland spar. In his Query 26, Newton asked, "Have not the Rays of Light several sides, endued with several original Properties?" This is a direct suggestion, but visualizing what might be an adequate shape for a light particle is quite difficult. Let us leave the issue in abeyance.

We go on to Observation 12, that light in passing through a thin layer of mica gives a beautiful – but puzzling – sequence of bright and dark rings. And light does the same when reflecting from the mica. How is this complex variation in transmission and reflection to be explained by the particle theory?

Rather than break our heads over this conundrum, let us turn directly to Newton and see how he dealt with the problem. First he proposed a certain microscopic property of light:

> Every Ray of Light in its passage through any refracting Surface is put into a certain transient Constitution or State, which in the progress of the Ray returns at equal Intervals, and disposes the Ray at every return to be easily transmitted through the next refracting Surface, and between the returns to be easily reflected by it.

Then Newton stated a definition:

> The returns of the disposition of any Ray to be reflected I will call its *Fits of easy Reflexion*, and those of its disposition to be transmitted its *Fits of easy Transmission*, and the space it passes between every return and the next return, the *Interval of its Fits*.

In Query 17 came an analogy and further explanation:

> If a stone be thrown into stagnating Water, the Waves excited thereby continue some time to arise in the place where the Stone fell into the Water, and are propagated from thence in concentrick Circles upon the Surface of the Water to great distances. And the Vibrations or Tremors excited in the Air by percussion, continue a little time to move from the place of percussion in concentrick Spheres to great distances. And in like manner, when a Ray of Light falls upon the Surface of any pellucid [that is, transparent] Body, and is there refracted or reflected, may not Waves of Vibrations, or Tremors, be thereby excited in the refracting or reflecting Medium at the point of Incidence, and continue to arise there ... and are not these Vibrations propagated from the point of Incidence to great distances? And do they not overtake the Rays of Light, and by overtaking them successively, do they not put them into the Fits of easy Reflexion and easy Transmission described above?

Query 29 capped this exposition with the thought that

> Nothing more is requisite for putting the Rays of Light into Fits of easy Reflexion and easy Transmission, than that they be small Bodies which by their attractive Powers, or some other Force, stir up Vibrations in what they act upon, which Vibrations being swifter than the Rays, overtake them successively, and agitate them so as by turns to increase and decrease their Velocities, and thereby put them into those Fits.

Newton has made a major addition to the particle theory: there is some kind of vibration of something that affects the light particles, making them alternately readily transmitted and readily reflected.

Indeed, Newton conjectured that there was an *ethereal medium*, a novel kind of fluid, that vibrated and produced the fits. This fluid was entirely distinct from the fluid formed when water molecules, say, condense into a liquid.

The entire addition – the fits and the ethereal fluid – is disturbing and unsatisfying. It introduces something foreign to the notion of little particles – light particles and atoms – and simple forces among them.

Why did Newton stick with light particles as a theory of light when they led to such an unsatisfying addition? Three reasons (at least) contributed to his decision.

First, Newton's era found the particle notion brilliantly successful in other contexts. The 1600s saw tremendous advances in chemistry. Gases and their properties were nicely explained by the hypothesis of atoms, another kind of little particle. Moreover, in his *Principia*, Newton had developed the equations that govern the motion of the planets around the sun and of the moon around the earth. These bodies are but particles on a large scale.

The particle notion was extremely fruitful; it might be the central concept to use in understanding the entire material world. In the long final query of the *Opticks*, Query 31, Newton spoke from his heart:

> All these things being consider'd, it seems probable to me, that God in the Beginning form'd Matter in solid, massy, hard, impenetrable, moveable Particles, of such Sizes and Figures, and with such other Properties, and in such Proportion to Space, as most conduced to the End for which he form'd them; and that these primitive Particles being Solids, are incomparably harder than any porous Bodies compounded of them; even so very hard, as never to wear or break in pieces; no ordinary Power being able to divide what God himself made one in the first Creation.

Second, in the ethereal medium Newton saw a link with gravity. In Query 21, he asked

> Is not this Medium much rarer within the dense Bodies of the Sun, Stars, Planets and Comets, than in the empty celestial Spaces between them? And in passing from them to great distances, doth it not grow denser and denser perpetually, and thereby cause the gravity of those great Bodies towards one another, and of their parts towards the Bodies; every Body endeavouring to go from the denser parts of the Medium towards the rarer?

The medium could do double duty: its vibrations could produce the "fits" of easy reflection and easy transmission, while its very presence and variation in density might account for gravity. That certainly is theoretical economy.

Third, Newton was convinced that nothing else would work. In his meticulous way, he had examined the other major theory of light that was current in his day and had found it wanting.

2.4 A decisive test

With his particle theory plus fits and an ether, Newton was able to explain well most of the optical phenomena known to him, and that was a wide range of phenomena, far wider than anyone else had ever investigated. In a few places, Newton was reduced to conjectures, but there was no behavior of light for which he did not have at least a plausible explanation.

Why, then, was I so negative in the final third of the last section?

Recall that Newton's analysis of refraction and ours, too, was based on a net *pull* toward water as light goes from air into water. Hence the speed of light in water must be larger than the speed in air. We inferred this in section 2.2.

No test of this proposition was available to Newton. The technology of his day did not permit it, or Newton, who was as able with his hands as with his mind, would have put the proposition to experimental test. Indeed, one must go forward in time 150 years before one finds the experiment. Prior to the period 1849–50, the speed of light had been measured by astronomical methods only, all yielding only the speed in vacuum.

The scene shifts to France. Léon Foucault and Hippolyte Fizeau – independently – strive to measure the speed of light in experiments conducted wholly on the earth's surface. Fizeau is the first to get a good value for the speed in air; the date is 1849. But Foucault wins the race to compare the speed in water with that in air; the time is 1850.

Figure 2.6 provides a schematic diagram of Foucault's experiment. We work our way through it in three stages.

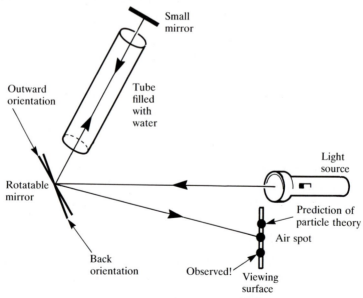

Figure 2.6 Foucault's method for comparing the speed of light in water with the speed in air. (The rendition is schematic only.)

Stage 1. Mentally remove the tube filled with water and hold the rotatable mirror fixed at the orientation labeled "outward orientation." Light travels leftward from the source, strikes the rotatable mirror (which we see edge-on), bounces up to the small mirror, is reflected straight back, bounces off the rotatable mirror again, and heads right back into the source. The "outward orientation" has been chosen so that the light beam will precisely retrace its path. The purpose of Stage 1 is to establish this.

Stage 2. Now we let the rotatable mirror rotate; its axis of rotation is perpendicular to the plane of the page. (We still mentally delete the tube filled with water.) Foucault used a miniature steam turbine to turn the mirror at a rate of 800 revolutions per second. Light comes from the source continuously, but only when the mirror is in the "outward orientation" will the reflected beam strike the small fixed mirror. Most of the time, the reflected beam hits the walls of the room and is lost. Light that does get to the small mirror takes about 3×10^{-8} seconds to make the trip from the rotating mirror up to the small mirror and back to the rotating mirror. That is a short time, but the mirror, turning at high speed, will have rotated a bit. When the light gets back to the mirror, the mirror has rotated to the "back orientation" and reflects the light off to the side, not straight back into the source. The light appears on the viewing surface at the "air spot," so called because only air is in the path of the light beam.

Stage 3. No longer do we mentally delete the tube filled with water. Rather, it stretches for 3 meters in the path light takes between the rotatable mirror and the fixed mirror. What change does Newton's particle theory predict?

The theory implies for light a higher speed in water than in air, and so light should take less time to go from rotatable mirror to small mirror and back. A shorter transit time implies that the mirror should rotate through a smaller angle. Thus the spot on the viewing surface should now fall between the air spot and the source. That prediction is unequivocal.

Equally unequivocal is the experimental result: the new spot falls *farther* from the source than the air spot. The prediction fails.

We can trace the failure back to its origin. For the spot to be farther away than in the air-only situation, the mirror must have turned through a larger angle. Since the turning occurs while the light is in transit to the small mirror and back, larger angle implies longer transit time. And longer transit time implies that the speed in water is actually less than the speed in air.

For Newton's particle theory, the outcome of Foucault's experiment is devastating.

We had asked, can we account for the behavior of light by supposing:

(a) that light is made up of little particles; and

(b) that those particles obey the same laws of physics as do other objects, for example, baseballs, whatever those laws may be in detail?

The answer is *no*. Whatever the nature of light may be, light is *not* like a tiny baseball.

2.5 Summary

Here is a capsule summary of Newton's particle theory.

- Light consists of little, hard particles.

- The light particles obey the same laws of motion that baseballs and planets do.

- Refraction is produced by a net force near the surface. On the transition air-to-water, the net force is a pull. The assumption of a net force plus the laws of motion yield Snell's law, a quantitative triumph for the theory. The analysis predicts that the speed of light in water is greater than the speed in air.

- Colors are explained by supposing that the light particles come in different sizes. The particles of red light are the largest; those of violet light, the smallest. This variation in size accounts for why the index of refraction depends on color. (The reasoning originally went the other way: from the variation in index with color to the hypothesis of different particle sizes for different colors.)

- Particles of light cannot be spherical but must have different sides – in some way not readily visualized. To account for the behavior of light when passing through Polaroid sheets, one needs such sidedness in the particles.

- To account for variations in the amount of light reflected and refracted, Newton assumed that the light particles could be put into "fits" of easy reflection and easy refraction. The conjectured cause of these fits is the interaction between light particles and a pervasive ethereal medium. Vibrations or waves in the medium spread out and, when they overtake a light particle, put it into one kind of fit or the other.

Foucault's experiment shows that a key prediction of this theory is wrong.

We will not try to patch up Newton's particle theory. Such a project would be hopeless. Rather, we note what we have learned about theory building in physics. We started with a simple central concept; here it was light as a particle, a particle akin to a tiny baseball. When experiment suggested it or seemed to require it, we added specific detail or embellishments; here it was different sizes or sides or fits and an ether. When we could, we tested the theory. Newton actually put his theory to many tests – successfully – that were not mentioned in our development; they fill pages of the *Opticks*. Rather, we leapt ahead to a decisive test – and saw that a magnificent edifice can collapse entirely.

Fortunately for physics, utter collapse is a rare occurrence. And, of course, our observations, amassed in chapter 1, are still valid.

The quotations from the *Opticks* have revealed some aspects of Newton the man. You may welcome more insight into this complex individual; the following section offers a view.

2.6 Newton himself

January 29, 1697. It had been a hard day at the office. Britain was calling in the old coins and issuing new ones. For the chief administrator at the Mint, this meant a lot of work: giving endless orders and keeping tabs on everything that should be done. And then to get home, expecting peace and quiet, and find this letter from the Continent. Johann Bernoulli poses a problem (as sketched in figure 2.7). Given are two points, labeled 1 and 2, the lower point displaced side-ways from the higher point. A wire connects the two points, and a bead can slide down the wire from one point to the other. Question: what shape should the wire have so that the bead, starting from rest, slides from point 1 to point 2 in the least amount of time?

There is no doubt about it. Bernoulli is out to embarrass him. Bernoulli supports Gottfried Leibnitz's claim to have invented calculus, and this mathematics problem – published in Europe and now sent privately to him – is supposed to show that only Leibnitz and his followers know the subject well enough to be able to solve such a "least time" problem. All right: challenge accepted.

The chief administrator sets to work, unaware that over the course of ten months only four persons on the continent of Europe will get to the point where they can even propose solutions: Johann and Jacob Bernoulli, the Marquis de L'Hospital, and Leibnitz. The midnight oil burns, but by 4 a.m. he is finished. He publishes his solution anonymously in the *Philosophical*

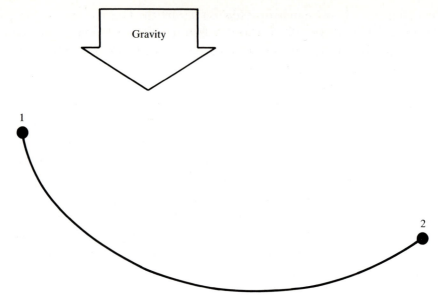

Figure 2.7 Johann Bernoulli's mathematics problem: find the shape of the curve joining points 1 and 2 such that a bead, starting from rest at point 1, will descend under gravity to point 2 in the least time. The shape shown here does provide the shortest trip and hence solves the problem. The steep drop at the start and the dip below point 2 enable the bead to pick up speed quickly. That shortens the time, despite the longer distance (relative to a straight line connecting the two points).

Transactions of the Royal Society for February 1697. There is no point in acknowledging publicly that he noticed the challenge.

When Bernoulli received a copy of the *Transactions*, he recognized the author (in Bernoulli's words) *tanquam ex ungue leonem*: just as, from the pawprint alone, one recognizes the lion. Chagrin could not suppress his admiration for Isaac Newton's brilliance. The problem was not merely solved. In the brief span of two small pages, it was solved with deftness and elegance.

At the time of this episode, Newton was fifty-five years old. He had become a civil servant just a year earlier; his title in the government was Warden of the Mint. That actually made him only second in command at the Mint, but his nominal boss did no work, and so Newton took charge. During the Great Recoinage, he was everywhere, doing everything. He studied the metallurgy of coining money and improved on it. He prosecuted counterfeiters and – since counterfeiting was then high treason – had them hanged. (Newton disliked that part of the job.) Under Newton, the Mint achieved an efficiency it had never seen before and that would be sorely missed – decades later – when he was no longer in charge.

This picture of Newton as an energetic, efficient administrator is a far cry from the popular image of Newton: a bemused old professor with curly white hair, in imminent danger of being hit on the head by a falling apple. Reality is stranger than myth.

Actually, there was an apple in Newton's life, but to find it we have to go to his youth. Newton was born in rural Lincolnshire in 1642, the year Galileo died. He grew up on a small farm – Woolsthorpe, it was called – and attended school in a nearby town. No sign of genius was apparent. Rather, Newton was just another bright lad, fond of making things with his hands and good at it, too. When a windmill was built nearby, he constructed a working model of it, just for his amusement.

The University of Cambridge was only sixty miles away, and Newton had the privilege of going to college. In today's language, he was a "work study" student. Although Newton had some money from home, he had to work to meet most of his expenses – and at quite menial tasks. He was eighteen when he entered Trinity College, one of the independent colleges that, taken together, made up the university, and soon he began to blossom. Isaac Barrow, professor of mathematics, took particular notice of young Newton and encouraged him.

In 1665, when bubonic plague once again spread through England, the university shut down and sent its students home. Dispersal was some protection against the rat-borne disease. Newton went back to the neighborhood of his childhood and spent the next eighteen months in intellectual isolation. For the young man of twenty-three, it was one of the two great periods in his life. Fooling around with the mathematics that he so enjoyed, he invented calculus, indeed, both differential and integral calculus.

He thought, too, about motion, both terrestrial and celestial. Following in Galileo's footsteps, Newton was grappling with the connections among force, inertia, velocity, and acceleration. For example, the moon moves around the earth (as we on earth see it) in an orbit that is nearly a circle. Some force must be bending the moon's trajectory into a circle. But what could be doing that? As Newton recalled fifty years later, he was sitting in the orchard one day, drinking tea and thinking about this puzzle, when an apple fell into the grass. Hmmm. Maybe the same force that pulls an apple to the ground pulls the moon's trajectory into a circle?

From astronomical data, Newton could figure out the size of the force needed to pull the lunar trajectory into a circular shape. If he supposed that the gravitational pull exerted by the earth decreases inversely as the square of the distance from the earth's center (for which he had some indirect evidence), then he could extrapolate from the earth's pull on an apple at the earth's surface to the earth's pull on the moon. When Newton made the numerical comparison – force needed versus force provided – he found that the numbers agreed, in his words, "pretty nearly."

The significance is not merely that Newton was getting things right. The deep significance lay in Newton's connecting celestial motion with terrestrial motion. For millennia, the stars, the planets, and the moon had been regarded as intrinsically different in their motion from apples and cannon balls, the kinds of things that move near the earth's surface. Now Newton was linking the motion of these disparate bodies. Two decades hence, Newton would go on to show that the same laws that govern the motion of apples govern also the motion of planets, their moons, and comets.

In 1667, Newton returned to Cambridge and began to move up the academic ladder at Trinity College. A year later, at the age of twenty-six, he became Lucasian Professor of Mathematics, a post he would hold for the next three decades. Richly endowed by Henry Lucas, the professorship had first been held by Isaac Barrow, Newton's mentor. When Barrow resigned it, he arranged for the young, substantially unknown Newton to be his successor. Newton's rise was as remarkable as it was swift.

In this chapter, we have learned some things about Newton's investigation of light and the theory to which it led him. Much of this work was done in the decade 1660–70, some at Cambridge, some at Woolsthorpe or thereabouts. Newton's study of refraction as a function of color led him to design and build a new kind of telescope. He reasoned that a lens of glass would never be able to focus all colors to the same image. Different colors are bent through different angles, and so they will not overlap perfectly in the image. Newton believed that this difficulty made the perfection of ordinary telescopes impossible or at least impractical. (The historical evidence is not clear on which inference Newton believed.) Newton's solution was to avoid refraction entirely: he would form the image with a curved mirror. Light of all colors reflects from a mirror through the same angle.

With his own hands, Newton built a reflecting telescope (8 inches long) that made objects seem forty times closer than they actually were. Along the way, he invented the alloy – of copper, tin, and arsenic – from which he polished the curved mirror. Newton was quite proud of his telescope.

Word of Newton's telescope made its way from Cambridge to the Royal Society in London and gave this body of scientifically inquisitive gentlemen its first indication that a quiet genius resided at Trinity College. (The group's full name was "The Royal Society of London for Improving Natural Knowledge." Formally organized in 1660 and chartered by King

Charles II in 1662, the Royal Society was devoted to "Experimental Philosophy," the inquiry into nature through experiment. In its early years, its meetings consisted primarily of recent experiments performed as lecture demonstrations and accompanied by general discussion. Among the Society's members in its first decades were Robert Boyle, Christopher Wren, Robert Hooke, and Samuel Pepys.) The Royal Society asked to see the telescope; Newton made a second telescope and sent it to them. In short order, Newton was elected a Fellow of the Royal Society (in January 1672), and his first scientific paper was read to a meeting of the society a month later. It described his extensive experiments on light and his demonstration – now seen as unequivocal – that white light is actually a mixture of all the colors of the rainbow.

Already in this early work we find the methodological differences that set Newton apart from his predecessors. Newton relied on experiment and observation, not on speculation from *a priori* philosophical principles. Accompanying this change was a shift in the kind of question asked of nature. Newton asked *how* light behaved and *how* objects moved; he did not ask *why* light behaved as it did or objects move as they do. Answering a "how" kind of question is difficult, but responses to the "why" kind are almost beyond reach. The understanding of nature that science has produced in the past three centuries comes in great measure from the shift in questioning that Newton initiated: the shift from "why" to "how." Someday, of course, we would like to understand "why" also, but only rarely is a deep answer to a "why" question possible.

Then, too, there is the comprehensiveness of Newton's investigations. Once Newton took up a subject, he looked at it from all directions. One experiment suggested another, and they were carried out with meticulous attention to quantitative detail.

But let us return to Newton's career at Cambridge. The decade of the 1670s saw Newton's interest switch from optics to chemistry and alchemy. It was not that Newton was avariciously trying to make gold out of lead. Rather, alchemy was a respected variant of what we now call chemistry, a variant in which "powers" were associated with various elements and chemical compounds, all in an attempt to unravel chemical behavior. No great success accompanied Newton's explorations here, and we can move on to 1684.

In that year, Edmond Halley dropped in to see Newton in Cambridge. He knew Newton – they had discussed comets two years earlier – and Halley thought that Newton might be able to answer a theoretical question that intrigued him: if an object – such as a planet – were attracted by a force that diminished with distance inversely as the square of the distance, what shape would the orbit have? Newton replied, an ellipse. That shape – a stretched circle – is also the empirical shape of planetary orbits. Johannes Kepler had discovered the elliptical character of the orbits while analyzing observations of the planets (around 1605), as both men well knew. Newton shuffled through his papers but could not find his derivation. (There is good reason to believe that he didn't look very hard. Ever since his first scientific paper had received severe – and unjustified – criticism, Newton had been reluctant to make anything public.) Halley, who numbered astronomy among his many pursuits, was keenly interested, and Newton promised to send him something.

What arrived was a nine-page manuscript containing a theory of motion, both terrestrial and celestial. Halley and the Royal Society immediately recognized the potential here and urged Newton to publish the brief work.

Thus began the three-year project that produced *Philosophiae Naturalis Principia Mathematica*, the Mathematical Principles of Natural Philosophy. The short manuscript became the merest of beginnings. Newton threw himself into the task of deepening and expanding. Only forty-two years old, he could work with vigor, and he did. The conceptual groundwork may have been laid at Woolsthorpe twenty years earlier, but there was much to be grappled with. Even the basic principles still needed amendment, and then there were a myriad of applications to be worked out and experiments to be done. In this second great period of his life, Newton became the proverbial forgetful professor. If a thought occurred to him at dinner when he had friends over, he excused himself from the table, went to write it down – and never came back, so immersed was he in the project.

In July 1687, the *Principia* emerged as one opus divided into three "books": the basic laws of motion, their application to fluids and to bodies moving in a fluid, and their application to the solar system. It was a stunning achievement, an intellectual tour de force such as the world had never seen. Though dauntingly difficult to read, the *Principia* sold well and immediately made Newton the foremost scientist of his age.

Figure 2.8 shows Newton at about this time in his life. The man who wrote the *Principia* was vigorous and intense.

Two years after the *Principia* appeared, Newton was elected by the university to be one of its two Members of Parliament. Religious matters were at issue in Parliament, and Newton was a deeply religious man. Indeed, theology was one of the four major pursuits in Newton's life. One might list them as follows: (1) physics and mathematics; (2) chemistry and alchemy; (3) administration of the Mint; and (4) theology and religious

Figure 2.8 Isaac Newton at age forty-six, two years after he finished the *Principia*. The locks are not a wig but rather Newton's own gray hair. At Newton's request, Godfrey Kneller painted the portrait in 1689. (Courtesy of Lord Portsmouth and the Trustees of the Portsmouth Estates.)

history. Newton was a Unitarian in an age when that was heresy and when public admission of such a faith would strip a person of his employment and perhaps more. Newton spent great blocks of time on his theological studies, first interspersed with his scientific work and then extensively in his old age. He was acknowledged as an expert on the history of the early Christian church, and several books on that subject and on Christian theology appeared in his old age or posthumously.

Around the time that Newton served in Parliament, he tired of his professorial life in bucolic Cambridge. (One must remember that Cambridge was hardly more than an overgrown village in those days. The university and the town offered Newton neither intellectual stimulation nor a congenial social life.) Seeking a change, Newton began to look for a government job in London. Yes, Newton actually sought an administrative job. The government owed him a favor for his support while in Parliament, but political plums did not fall as readily as apples; for some years, Newton despaired that his friends would be able to find him a position. In 1693, as a consequence of years of intellectual effort, of disappointment in his job search, and perhaps of other reasons, Newton came close to a nervous breakdown. At last, in 1696, he was appointed Warden of the Mint and moved to London. The transition takes us to the point where this biographical sketch started.

Newton proved to be an excellent administrator. He was promoted to the top job – Master of the Mint – when the incumbent died, and Newton retained that job until his own death, at age eighty-four, in 1727. In 1703, he was elected President of the Royal Society and exercised a firm hand in directing its affairs, again until his death.

Earlier in this chapter we noted that Newton published his *Opticks* in 1704, immediately after Hooke's death. Substantially all his scientific achievements were then public. So profound and so numerous were they that any era would have been proud to have Newton among its members. And royalty rose to the occasion: in 1705, Queen Anne honored Newton with the first knighthood awarded for scientific achievement. The farm lad from Woolsthorpe was henceforth Sir Isaac Newton.

To be sure, party politics played an essential role in Newton's knighthood, too. Newton had served in Parliament twice (in 1689 and 1701), but he had not run in the most recent election. Charles Montague, the politician and friend who had gotten Newton his job as Warden of the Mint, sorely wanted Newton to run in 1705. Newton acceded and waged a vigorous campaign among the faculty of the university (for he was running to represent the University of Cambridge). Montague arranged for Queen

Anne to provide Newton a "great Assistance" (in Montague's words) in his campaign: the knighthood, bestowed on Newton where it would make the greatest impression – Trinity College, Cambridge.

Despite the royal recognition, Newton lost. In the election, religious issues seethed just below the surface and even erupted in Cambridge. Running for public office threatened to expose Newton's Unitarian religious views, something he could not afford to have happen, and Newton never again ran for Parliament. We can return from politics to loftier matters.

British science is full of titles from the nobility: Sir Arthur Eddington, Dame Kathleen Lonsdale, Lord Kelvin, Baron Rayleigh, and a nearly endless list of others. When I was an undergraduate, I was greatly impressed. Boy, I thought, the British nobility sure has produced a lot of great scientists. Only later did I realize that the logic flows the other way: as with Newton, commoners are elevated to the nobility for their scientific work. Although the practice started with Newton in 1705, it remained a rare occurrence until the 1830s.

This biographical sketch draws now to a close. There is so much more that one could write about Newton – for example, his personality and his friends – but that will have to be left to the biographies noted among the Additional resources for this chapter. A particularly delightful short biography is Edward Neville da Costa Andrade's *Sir Isaac Newton*. A quotation from it concludes this section.

> Newton could often speak very modestly of his own achievements. Shortly before his death, he said, "I do not know what I may appear to the world; but to myself I seem to have been only like a boy, playing on the seashore, and diverting myself in now and then finding a smoother pebble or a prettier shell than ordinary, while the great ocean of truth lay all undiscovered before me." We can explain this, perhaps, by supposing that Newton's aim was to understand the whole scheme and mystery of the world, in which he believed that the truths of exact science were only a part of a greater truth.

Additional resources

E. N. da C. Andrade's biography, *Sir Isaac Newton* (Collins, London, 1954), captures the essence of Newton and his achievements in 140 pages. It makes delightful reading. Though out of print at the time I write this, the book is in the public domain and may be photocopied at will.

Richard Westfall provides a modern, extensive biography in *Never At Rest: A Biography of Isaac Newton* (Cambridge University Press, New York, 1980). The book is available in paperback. The writing flows smoothly, and one fascinating anecdote follows another.

A modern appreciation of Newton is provided by *Newton's Dream*, edited by Marcia Sweet Stayer (McGill–Queen's University Press, Montreal, 1988). Eight speakers at a symposium commemorating the *Principia*'s tercentenary write about facets of Newton and his work.

Many additional references can be found in the Westfall and Stayer volumes. And one should not overlook Newton's *Opticks*, available as a paperback from Dover Publications (New York, 1952).

Questions

1. The speed of light as measured in a certain substance is 2×10^8 meters/second.

(*a*) According to the particle theory of light, what should be the index of refraction of this substance? Take the speed of light in vacuum to be 3×10^8 meters/second and the speed in air to be essentially the same; the speed in air differs from the speed in vacuum by less than 1 per cent, and so small a difference you may ignore here.

(*b*) What would be the path of a light ray which is incident from air onto the substance at an angle of 60 degrees from the normal to the surface? (Do use Newton's particle theory of light, but be careful.) Figure 2.9 shows the proportions among the sides of a 60° right triangle.

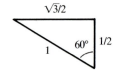

Figure 2.9 Note that $\sqrt{3}=1.73$.

(*c*) Part (*b*) of this question is subtle, but it can lead to a neat test of the particle theory. Suppose you did have some of the "certain substance." Can you devise a simple test of the particle theory of light? How?

2. From time to time we need to work with quantities, such a force or a velocity, that have both a *direction* and an *amount* (a *magnitude*) associated

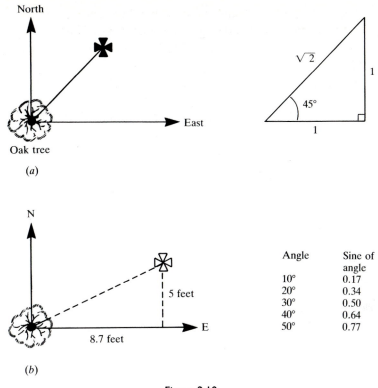

Figure 2.10

with them. Here are two illustrations of how such a quantity can be represented by components.

(a) Part A of figure 2.10 depicts a map on ancient vellum. To find the buried treasure, walk 14 feet northeasterly from the oak tree. If the pirate would prefer to go first due east and then due north, how far should the pirate go in each of those two directions? (For reference, $\sqrt{2}=1.4$.)

(b) On a different island, as noted in part B of the figure, the treasure map indicates that gold is to be found by going east 8.7 feet and then north 5 feet from another gnarled oak. If one walks in a straight line from this oak, how far must one go? Then figure out the angle with respect to due east that one's path should make.

3. If you have read Andrade's biography of Newton, here are several questions.

(a) Andrade distinguishes sharply between Newton's style in natural

57

philosophy and that of his predecessors. Write a paragraph describing that distinction.

Why is Galileo an exception?

(*b*) What does Andrade see as the significance of Newton's prism experiments? Do you concur?

(*c*) Andrade attempts to summarize Newton's accomplishments in optics. Outline those accomplishments (taking for your reader a fellow student in your course).

3 A wave theory of light

I looked toward the cliffs once more and the wave-tossed miles in between. "Let's go," I said, and without more ado we headed out into the open.

Sigurd Olson,
The Lonely Land

3.1 Waves

You may have arrived here – the start of chapter 3 – by having read through chapters 1 and 2, or you may have jumped directly from the observations in chapter 1 to this chapter. If the latter is the route you took, let me note the central conclusion of chapter 2: we can *not* build a successful theory of light by supposing that light consists of little particles which obey the same laws of motion that baseballs and planets do. Whatever the nature of light may be, light is *not* like a tiny baseball.

We need to try a radically different theory of light.

A particle was appealing because it travels through space, as does light. What else travels through space?

Figure 3.1 provides a specific answer. We take a 5-meter length of rubber tubing, the kind used to carry gas to a Bunsen burner, and tie one end to a firm support. We stretch that hose moderately tautly and then strike it sharply from below with a stick. The blow deforms the hose, producing an upward bulge. And then the distortion moves along the hose: we see a

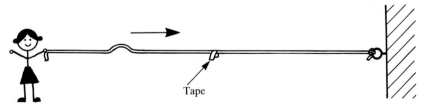

Tape

Figure 3.1 A shape traveling along a stretched length of rubber tubing. When the shape, traveling to the right, reaches the rectangular piece of tape, the hose so marked will move up and then back down. Thus, as the shape passes, the tape and hose move perpendicular to the direction in which the shape moves.

shape traveling through space from left to right. If we put a piece of tape somewhere along the hose, so that we can focus our attention on a specific bit of hose, we notice that the rubber particles do not travel along the hose. Rather, when the traveling shape reaches them, the rubber particles move at right angles to the hose.

When the moving shape gets to the fixed end of the hose, the shape turns over and heads back to us. A reflection of sorts occurs. Like light from a mirror, perhaps?

Shapes formed in other ways move through space. When a stone is tossed into a pond, the ripples on the water's surface move outward in concentric rings. A storm generates huge ocean waves; they roll on, mile after mile.

What we have here are three instances of *wave motion*. Just what the essence of a "wave" is, is surprisingly hard to state. Here is how I try:

- A wave is a *pattern*, not a thing (like a ball).
- Usually a wave is a moving, changing pattern.

Of course, the pattern must be formed out of something (for example, out of a rubber hose or the surface of water), but that "something" need not move in the same way that the wave does. Succinctly, there must be "something" of which the pattern is formed. But let us leave this attempt at abstraction and just learn more about waves. After all, we recognize a wave when we see one.

In chapter 2, we had an idea for how to develop a successful theory of light. We pursued the idea but failed to reach our goal. Now we need a new plan. Here it is, with some results revealed:

(*a*) Learn about waves.

(*b*) Find that light does indeed have a wave-like character.

(*c*) Learn what the "something" is out of which light waves are formed.

Parts (*a*) and (*b*) occupy us in this chapter and the next. Chapter 5 addresses part (*c*).

3.2 Some general properties of waves

For showing waves, a rubber hose is a good start, but another piece of apparatus will do even better. It is rather like the fish skeleton that appears in comic strips. Figure 3.2 shows three views. The ribs are metal rods, 4 millimeters in diameter and 20 centimeters long. They are attached firmly to

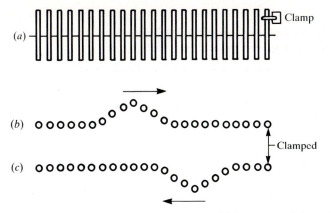

Figure 3.2 An apparatus for demonstrating twist waves. (a) Top view of the rods and the fiber. (b) Side view, showing just the ends of the rods on the side nearer to us. The hump is a "crest." (c) The shape in part (b) has reflected from the clamped end; it returns as a depression: a "trough." The arrows adjacent to the shapes indicate the corresponding velocities.

a long fiber of flexible steel. The fiber is supported horizontally 15 centimeters above a table top, amply high enough for the ends of the rods to swing up and down. When one end of a rod is moved upward, say, the central fiber is twisted (along its length). That twist causes nearby rods to begin to move upward, too, and the process propagates along the skeleton. But we do not really need to know how the wave motion is produced; it is enough for us to try various things and see how waves actually behave.

We start with all rods horizontal and note that the rod at the far right end is firmly clamped in place. Snappily, we lift the rod at the left end and then return it to its original position. A hump, which we will call a *crest*, moves away to the right; that crest is the wave, moving horizontally.

The pattern – the crest – is formed by the ends of the metal rods. Those rods are the "something" out of which the wave is formed. The rods twist up and down, and so the motion of the "something" is perpendicular to the motion of the wave.

When the crest reaches the clamped end of the skeleton, a new wave is sent back: a depression, which we will call a *trough*, moving to the left. Thus we get a kind of reflection.

Variation in speed

If we use a skeleton with shorter rods, 10 centimeters rather than 20, we still get waves, of course, but a crest now moves much more rapidly. We are still exploring twist waves, but the rod length is numerically different, and so is the speed. Thus we find a change in the wave speed when we make a quantitative change in the context. This change is akin to the change in the speed of light when light passes from water to air.

Partial transmission and partial reflection

We can connect two skeletons together, one with short ribs, the other with long, as illustrated in figure 3.3. If we start a crest at the left-hand end, it moves swiftly toward the right. When the crest encounters the boundary between the two skeletons, a crest – reduced in height – is transmitted into the second skeleton. And a trough returns leftward in the first skeleton. Thus we find partial transmission and partial reflection when the wave is incident upon the interface between two regions where the wave speeds differ. This reminds us of the behavior of light incident from air onto water: some light is transmitted (and perhaps refracted), and some is reflected.

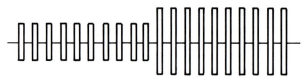

Figure 3.3 Two skeletons connected so that, in effect, there is only a single long central fiber.

Superposition

Now we return to just the long-ribbed skeleton; waves move more slowly on it, and we can better see what is happening. From each end, we send a crest toward the center. Part (a) of figure 3.4 shows this, and parts (b) and (c) show the situation when the crests coincide and then somewhat later. The neat question is this: do the crests bounce off each other, or do they pass

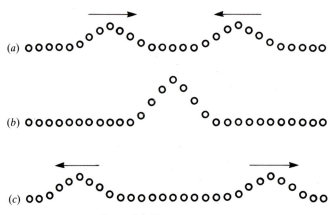

Figure 3.4 Two crests meeting.

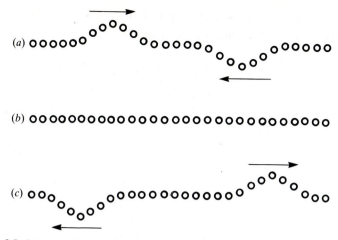

Figure 3.5 A crest and a trough meeting. In part (b), all the rods are instantaneously horizontal, but some are only passing through the horizontal orientation. Those rods just to the right of center are rising and will shortly form the crest; those just to the left are descending and will form the trough.

through each other? Because the crests are identical in size and shape, we cannot tell which alternative is correct.

To keep track of the waves, we form one as a crest and the other as a trough. Figure 3.5 shows what now transpires. The evidence is unmistakable: the waves pass through each other. This, of course, is akin to light beams that intersect without perceptibly affecting each other.

Moreover, we meet for the first time a central property of waves. The large crest in part (b) of figure 3.4 looks like what one would get by adding together the two crests that overlap there. Indeed, the large crest was constructed that way, and it reproduces faithfully what small crests actually do. The flat middle section of figure 3.5(b) is what one would get by adding algebraically a crest and a trough. For example, a trough of 2 centimeters' depth would be counted as a negative value, −2 centimeters, to be added to a crest of +2 centimeters, thus yielding zero and hence a flat shape.

What we note here is often codified as the *superposition principle*:

> to get the resultant pattern during overlap, add the original patterns algebraically.

The word "superposition" is used in its sense of "to superimpose" or "to lay one on top of the other." That is what we do, and then combine algebraically, so that a trough cancels out a crest.

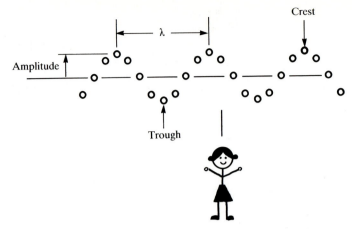

Figure 3.6 A periodic wave. Crests and troughs, generated off-page at the left, move toward the right. (To prevent reflected waves from destroying the simplicity of this periodic wave, one needs a skeleton infinitely long or a way to simulate that ideal situation; a dash pot at the right-hand end will do nicely.)

Periodic waves

Isolated crests and troughs are not the whole story. If we move the rod at the skeleton's left end up and down periodically, we generate a sequence of crests and troughs and so produce a periodic wave. Figure 3.6 illustrates this.

The distance between adjacent crests is called the *wavelength*. It is denoted by the lower case Greek letter lambda, λ, chosen (presumably) as a mnemonic for "length."

If we move the end rod up and down three times a second, we generate three new crests each second (and three new troughs). Alice, stationed off to the right, watches these crests go by her. The shape oscillates from crest to trough and back to crest three times each second. The number of oscillations per second (at a fixed location) is called the *frequency* of the oscillation or periodic wave. The symbol that we will use for frequency is simply f.

A tight connection links the frequency f, the wavelength λ, and the speed v of a periodic wave. Figure 3.7(a) shows a crest about to pass our observer. If we focus our attention on that specific crest as time goes on and as the crest moves to the right, we can tell where the crest is 1 second later, as indicated in part (b). In our case, a frequency of three oscillations per second tells us that the specific crest and two others have passed Alice. The specific crest now lies three wavelengths off to the right.

The wave speed v is the ratio of the distance traveled by a crest to the time elapsed. So we may write

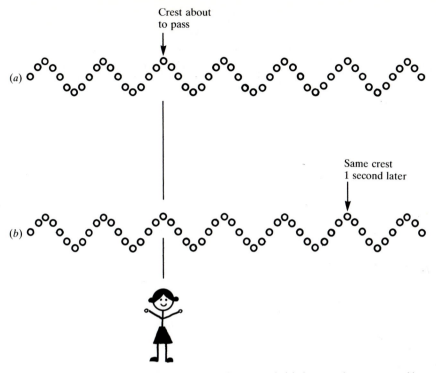

Figure 3.7 Relating frequency and wavelength to speed. (*a*) A crest about to pass Alice, our observer. (*b*) The situation 1 second later.

$$\text{wave speed} = v = \frac{\text{distance travelled by crest in 1 second}}{1 \text{ second}}$$

$$= \left(\begin{array}{c} \text{number of crests} \\ \text{that pass per second} \end{array} \right) \times \left(\begin{array}{c} \text{distance} \\ \text{between crests} \end{array} \right)$$

$$= f\lambda.$$

For each crest that passes, our specific crest finds itself one wavelength farther to the right. To find out how far it goes in 1 second, we multiply the number of crests that pass by the wavelength.

In short, the link among speed, frequency, and wavelength is this:

$$v = f\lambda.$$

If you need this relationship sometime and are no longer sure how the pieces should be put together, you can reason it out by thinking about the units. Wavelength has the units of length: meters. Frequency has the units of number of oscillations per second; "oscillations" does not count as a unit, and so frequency has units of "per second" or one over seconds. A speed must have the units of length over time, that is, meters per second. The only way to combine λ and f to get the units that v must have is by multiplying λ and f: meters times "per second" gives meters per second.

The height of a crest above the mid-line of the wave is called the *amplitude*. Basically, the amplitude tells us how big the wave is (in a precise fashion).

3.3 Light as a wave?

We turn to theory building again, and we ask, can we account for the behavior of light by supposing that light is wave motion of some kind?

To begin to answer the question, we need to compare the way that waves reflect and refract with the way that light does.

So far, the waves we have studied have moved along a line: along the skeleton, say. (To be sure, the rods moved perpendicular to the fiber, but the wave itself – the moving shape – moved along the fiber, to the right or to the left.) Light rays can impinge on an interface obliquely, and so we need to extend our study of waves to two dimensions at least.

Ripples on a pond are an example of waves moving in two dimensions: the surface of the pond. Indeed, water waves provide an excellent system to study, and figure 3.8 shows a good way to go about it. The water is in a large

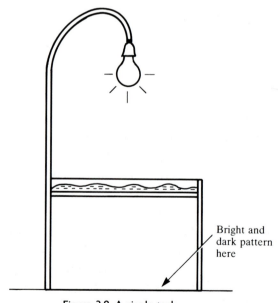

Bright and
dark pattern
here

Figure 3.8 A ripple tank.

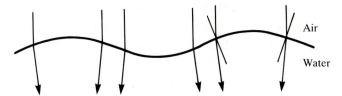

Figure 3.9 How the ripples affect light rays. For the two rays at the right-hand end, the sketch shows the normal to the surface where the rays impinge. On the transition air-to-water, the ray is bent toward the normal, giving the refracted rays shown there. A crest concentrates the light under itself, and a trough spreads out the light.

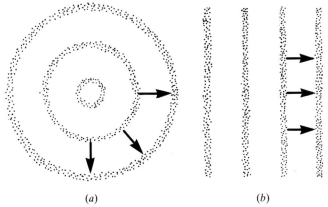

(a)   (b)

Figure 3.10 Two patterns below a ripple tank. A dense set of dots represents the concentration of light by a wave crest. Troughs are not explicitly delineated; they lie between adjacent crests. (a) A finger dipped into the water produces concentric rings of crests and troughs. (b) A long board moved up and down periodically produces a periodic wave with straight crests and troughs. We will speak of the crests in a single ring or a single straight line as forming a single crest line.

tray with a clear glass bottom. A light bulb provides illumination, sending light through the water and onto a sheet of white paper laid on the floor. How do the ripples reveal themselves? Figure 3.9 shows how the ripples refract the light, concentrating it under the crests and leaving the trough regions dark by default. Thus a pattern of crests and troughs on the water's surface is transformed into a similar pattern of bright and dark regions below the ripple tank.

Figure 3.10 illustrates the patterns produced under the ripple tank in two situations. The arrows indicate the velocities of the corresponding waves. The rings expand outward, and so their motion – their velocity – is radially outward. For each portion of curved crest line, the wave velocity is at right angles to that portion. In part (b), the long straight crest lines move to the right, and their velocity is also perpendicular to the crest lines themselves.

When looking at ripple tank patterns, we must remember this perpendicularity. The motion of a crest line is at right angles to the crest itself. (This perpendicularity is different from the perpendicularity we noted in the twist waves; both are vital relationships.)

Reflection

Now we can use the ripple tank to answer experimentally two critical questions about waves.

Question 1. Do waves reflect the way light does?

Figure 3.11 shows the pattern produced by periodic water waves striking a tall barrier and reflecting. (You may be fortunate enough to see such reflections in real life or via a film loop.) Reflection is evident, and the symmetry of the situation assures us that the angle of reflection equals the angle of incidence. To Question 1, the answer is certainly "yes."

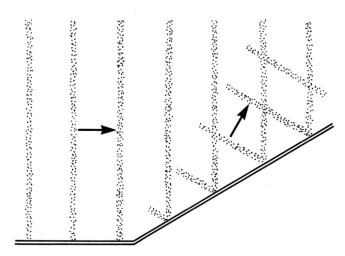

Figure 3.11 Waves reflecting from a barrier placed obliquely in their path. We are looking straight down on the scene. The reflected crests lie diagonally up and to the left, but they move diagonally up and to the right. The length of each reflected crest line is equal to the length of that portion of the incident crest line which has struck the barrier. That is why the reflected crest lines grow.

Variation in speed with depth

To simulate refraction with water waves, we need two regions with different speeds for the water waves. Just changing the depth of the water – by placing a thick piece of clear glass on the bottom, as in figure 3.12 – will do the trick. Waves move more slowly in shallow water.

When a periodic wave makes the transition from deep to shallow water,

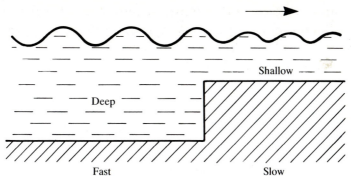

Figure 3.12 A side view of how depth affects the speed of water waves. Here the waves are headed directly toward the step from deep to shallow water. (The amplitude of the transmitted wave (on the right) is less than the wave amplitude on the left because a portion of the incident wave is reflected.)

its speed decreases. We learned that the speed is equal to the product of frequency and wavelength: $v=f\lambda$. Thus the frequency or the wavelength must change – or perhaps both of them. We can reason out what must happen. If the wave in deep water has a frequency of three oscillations per second, then three crests arrive at the deep-to-shallow boundary each second. Crests cannot accumulate at the boundary, nor can any brand-new ones be produced. So, each second, three crests must move away from the boundary into the region of shallow water. In short, the frequency of the periodic wave stays constant during the transition.

Because the speed decreases during the transition but the frequency remains constant, the wavelength must change and must decrease. We could have figured that out in another way, as follows. A crest moves more slowly in the shallow water. After it passes the boundary, the crest moves less distance before the next crest gets to the boundary than it would have moved in equal time in the deep water. No wonder the wavelength is shorter.

Refraction qualitatively

Our next question is this:

Question 2. Do waves refract the way light does?

Figure 3.13 provides a response. A periodic wave with straight crest lines moves in from the left. At the boundary, each crest line develops a kink because part of the crest line slows down while the remainder continues at the deep-water speed. The direction in which a crest line moves is perpendicular to the crest line itself. Because the crest lines are tilted in the new

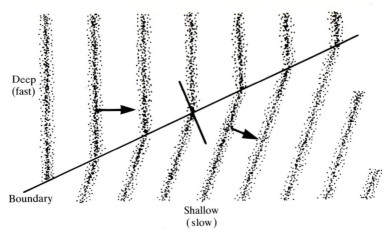

Deep
(fast)

Boundary

Shallow
(slow)

Figure 3.13 Waves refracting on the passage from deep to shallow water. For the speed, the transition is from fast to slow. The arrows indicate the wave velocity; the short line perpendicular to the boundary is the normal. Some reflection occurs, too; for clarity's sake, the reflected waves have been omitted.

region, the direction of wave motion changes, and so refraction certainly occurs.

Moreover, we can compare the directions in which the wave is moving – before and after passage through the boundary – with the normal to the boundary. To do this, we focus attention on the velocity arrows and their relation to the normal. We find that the wave is bent toward the normal. The transition here is from fast to slow speed, and we find the same sense of bending as when light makes the transition from air to water, where – thanks to Léon Foucault – we know that the transition is from fast to slow speed for light. This is auspicious.

Qualitatively, the answer to Question 2 is "yes." But what about the details, the quantitative behavior? Does a version of Snell's law hold for waves?

3.4 Refraction quantitatively

We can build on what figure 3.13 has shown us. For graphic clarity, let us compress into a single line the stipples that previously represented a crest line. And we shift the apparatus around so that the boundary line runs horizontally on the page, to make things similar to the diagrams in chapters 1 and 2. Then figure 3.14 displays two adjacent crest lines of a periodic wave. Each line has a kink where the crest line meets the boundary between

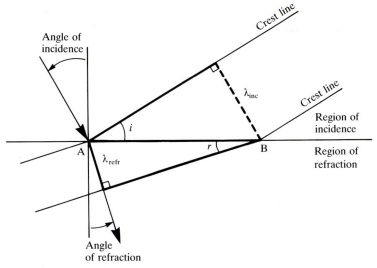

Figure 3.14 The quantitative analysis of refraction. Two adjacent crests appear in the regions of incidence and refraction. The heavier lines outline the two essential triangles.

the region of incidence and the region of refraction. As before, the wave speed is less in the region of refraction. The separation of the crests is always one wavelength, but the very length of one wavelength differs between the two regions; we learned that in the preceding section of this chapter. Thus the separations of the crests are denoted by λ_{inc} and λ_{refr}, for the "incident" and "refracted" waves. The straight arrow in the upper left-hand quadrant is perpendicular to the crest line in the region of incidence and shows the wave's direction of motion as the wave approaches the boundary. The angle between that arrow and the normal is the angle of incidence. The straight arrow in the lower right-hand quadrant serves the same function for the refracted wave. The arrow is perpendicular to the crest line in the region of refraction, and the angle between the arrow and the downward extension of the normal is the angle of refraction.

To see whether waves show a behavior identical to Snell's law for light, we need to compute a ratio of sines. This requires four steps.

First, we establish that the angle marked simply i is equal to the angle of incidence. The unmarked angle between those two angles, when added to either angle i or the angle of incidence, gives a right angle. So angle i and the angle of incidence must be equal.

Second, we prove that the angle marked simply r is equal to the angle of refraction for a similar reason. The angle r, the length AB (along the boundary), and the length λ_{refr} form a right triangle. The unmarked acute angle in that triangle plus the angle r must equal 90 degrees. Moreover, that

71

unmarked angle and the angle of refraction equal 90 degrees. Hence the angle r and the angle of refraction are equal.

Third, we note another useful right triangle, this one formed by the angle i, the length AB, and the length λ_{inc}.

Fourth, we use those two triangles to work out the ratio of sines, as follows:

$$\frac{\sin(\text{angle of incidence})}{\sin(\text{angle of refraction})} = \frac{\dfrac{\lambda_{inc}}{\text{length AB}}}{\dfrac{\lambda_{refr}}{\text{length AB}}}$$

$$= \frac{\lambda_{inc}}{\lambda_{refr}}$$

$$= \frac{f\,\lambda_{inc}}{f\,\lambda_{refr}}$$

$$= \frac{v_{\text{in region of incident wave}}}{v_{\text{in region of refracted wave}}}. \qquad (1)$$

For the first equality, we use the two triangles to express the sines. Then we multiply numerator and denominator by "length AB" and cancel that length out; we find just a ratio of wavelengths. The frequency of the periodic wave remains constant through the transition; so we may multiply numerator and denominator by the frequency f. Then our general result for periodic waves, $v = f\lambda$, gives us the last line.

We find that the ratio of sines is equal to a ratio of speeds, and the speeds, of course, are characteristic of the regions. The ratio of sines does not depend on the angle of incidence. The expression in equation (1) has the structure of Snell's law, but we were misled once, in chapter 2. Let's not jump to a hasty conclusion.

Comparison with light

For a comparison with the actual behavior of light, we need to adapt equation (1) so that it applies – tentatively – to light. We know how light refracts when it makes the transition from air to water, and so we take the region of incidence to be air and the region of refraction to be water. Thus, if light is a wave of some kind, then equation (1) implies

$$\frac{\sin(\text{angle of incidence})}{\sin(\text{angle of refraction})} = \frac{v_{air}}{v_{water}}. \qquad (2)$$

Section 2.4 described Léon Foucault's experiment of 1850. Foucault established unambiguously that v_{water} is less than v_{air}, but he never published a value for v_{water} or a numerical ratio of speeds. Three decades later, Albert Michelson, professor of physics at the Case Institute in Cleveland, Ohio, repeated Foucault's experiment. In a report dated August 1883, Michelson noted that he had made six independent determinations of the ratio v_{air}/v_{water} and that the average value was 1.33:

$$\frac{v_{air}}{v_{water}} = 1.33.$$

The experimental uncertainty was less than 1 per cent; so Michelson could assert with confidence that light moves 33 per cent more swiftly in air than in water.

When we insert Michelson's numerical value for v_{air}/v_{water} into equation (2), we find

$$\frac{\sin (\text{angle of incidence})}{\sin (\text{angle of refraction})} = 1.33.$$

A glance back to section 1.1 reminds us that the index of refraction of water is indeed 1.33. The agreement is excellent.

Actually, we should pause and ask just what "agreement" signifies here. The number 1.33 in the original Snell's law of chapter 1 came from angle measurements. To be sure, we did not make the measurements, but others carefully measured angles of incidence and refraction, computed a ratio of sines, and found the number 1.33.

The number 1.33 of the present chapter comes from speed measurements, an entirely different origin.

Indeed, figure 3.15 displays the logic of the test, which goes like this. If

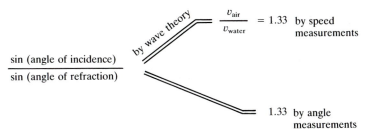

Figure 3.15 The logic of the test. The physical context is light passing from air into water. Speed measurements and the wave theory predict that the ratio of sines will equal 1.33. Angle measurements confirm that prediction.

light is a wave of some kind, then the ratio of sines must be equal to a ratio of speeds, as derived in this section. Consequently, the measured ratio of speeds enables us to *predict* a ratio of sines equal to 1.33. Because the *measured* ratio of sines is actually 1.33, the prediction is confirmed. Refraction, considered quantitatively, provides a *bona fide* test for a wave theory of light – and the theory passes superbly!

Mnemonic

Wave motion in two dimensions is difficult. What we have just worked our way through requires concentration, and in all the small steps, it is easy to lose sight of the big picture. A mnemonic for the process of wave refraction would be nice to have.

I grew up in Massachusetts, a state that remembers the Revolutionary War as though it had happened yesterday. The British soldiers who marched to Concord on the 19th of April in 1775 are still remembered, especially their retreat. They provide the nominal context for the mnemonic sketched in figure 3.16. The soldiers march in straight rows so long as they are on the land. Then, as some soldiers cross the boundary into a marsh, they slog along more slowly. Each row develops a kink that separates the soldiers still on land from those splashing through the marsh. Of course, as well-disciplined soldiers, the Red Coats try to maintain straight lines, but inevitably there are two sets of lines: one on land and the other in the

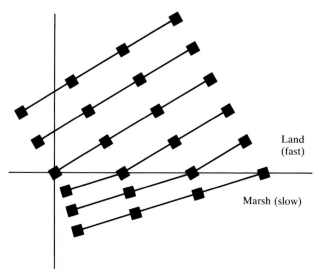

Figure 3.16 A mnemonic for refraction by waves. We see the British soldiers, each represented by a square, from directly above the terrain.

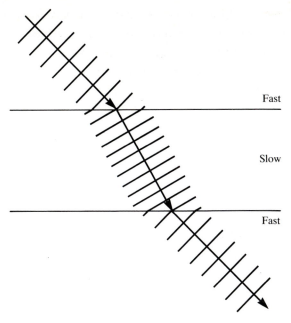

Fast

Slow

Fast

Figure 3.17 A wave making two transitions, the boundaries being parallel to each other. The wave speed is slower in the middle region than above or below. The three arrows are drawn perpendicular to the crest lines; thus they indicate the direction in which the crest lines move. If we think of a "ray" of light, the arrows indicate also the path of a ray of light.

marsh. The reduction in the soldiers' speed refracts the troop column toward the normal.

This section closes with a final sketch that ties together our understanding of refraction. We may interpret figure 3.17 in either of two ways. We may look at it as showing water waves moving from deep to shallow water and then back to deep water. Or we may interpret the sketch – tentatively – as showing a light wave going from air into glass and then back to air. In either interpretation, the wave goes from a region of high speed to one of low speed and then back to a region of the same high speed. The arrows drawn perpendicular to the crest lines show the direction in which the crests move. We see first bending toward the normal and then away from it.

The figure provides also an opportunity to make a point well worth emphasizing. Previously, we noted that, in a ripple tank and elsewhere, the motion of a crest line is at right angles to the crest line itself. If we think – for a moment – of light in terms of "rays" of light, then figure 3.17 shows us also the path taken by a ray of light. It is the sequence of three arrows. The ray's path is perpendicular to the crest lines. But of course – in the sense that a "ray" of light and the crest lines that represent a light wave must move in the same direction. Yet this is a point easily overlooked or misunderstood, and so it is worth repeating here: the path of a "ray" of light is perpendicular to the crest lines of the light wave.

3.5 Waves in perspective

It is worth our while to recapitulate the essential points of this chapter. Here they are.

- A wave is a pattern, not a thing.
- The pattern, of course, must be formed out of something.

In this chapter, we studied waves on a rubber hose, on a metal skeleton, and on the surface of water. In all three of these contexts, the motion of the "something" is at right angles to the motion of the wave. Such waves are called *transverse* waves. They are not, however, the only kind of wave, in a geometric sense. The next paragraph describes another kind.

Imagine yourself standing in line, waiting to buy tickets for a popular movie. Perhaps you are tenth in a queue of thirty persons. The couple at the ticket window complete their purchase and step off to the side. The person behind them steps forward; then the next person steps forward; and soon you, too, step forward one pace. A "crest" of "one step forward" sweeps down the queue, from the ticket window to the end of the line. But you and everyone else take one step forward. Your motion and the crest's motion are along the same line, and so such a wave is called a *longitudinal* wave. In this chapter, we concentrated on transverse waves because they are far more significant for our ultimate purpose: understanding light.

If we change the length of a skeleton's ribs or the depth of the water in a ripple tank, the speed of the waves changes. Thus, if we make a quantitative change in the context, the wave speed changes.

When a wave impinges on the boundary between regions with different wave speed, a portion of the wave is transmitted and the remainder is reflected.

When this process occurs in two dimensions (or three), the angle of reflection equals the angle of incidence. The direction of the transmitted wave is determined by the ratio of sines and the ratio of speeds, as derived in equation (1) of section 3.4.

When two (or more) waves overlap, we can figure out what the net result will be by invoking the superposition principle: to get the resultant pattern during overlap, add the original patterns algebraically.

Periodic waves are characterized by their wavelength λ, their frequency f, and their amplitude. The wave speed v, the wavelength, and the frequency are linked: $v=f\lambda$. If we know any two of these quantities, we can compute the third.

When a periodic wave strikes an interface, the transmitted wave will have the same frequency as the incident wave.

Finally, if we suppose that light is wave motion of some kind, then we find that our nascent theory successfully describes the reflection and refraction of light. That is, given a specific direction of incidence and given measured speeds of light, the wave theory correctly predicts the directions of the reflected and refracted light.

To be sure, we have not addressed the question, what is the "something" out of which a light wave is formed? And even the very idea that light is some kind of wave motion would benefit from further experimental test. In the next chapter, we work on the second of these items.

Additional resources

Film loops on "Wave Motion in a Ripple Tank" may be purchased from Kalmia Company, 21 West Circle, Concord, MA 01742. They are a welcome adjunct to a real ripple tank and its technical difficulties. In short, the films and the original complement each other well.

Questions

1. (*a*) Figure 3.18 provides a sketch (drawn to scale) of two crests on a rubber hose at a certain instant of time. The crests move toward each other; their speed is $v=500$ centimeters/second. What shape does the hose have after 1/100 of a second has elapsed? After 2/100? After 3/100? Please sketch the shapes.

(*b*) Now suppose the shape coming from the right is inverted, is "down-

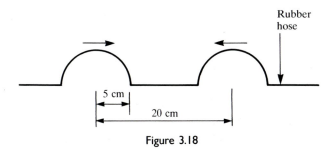

Figure 3.18

stairs, not upstairs," and hence is a trough. What shape does the hose have at the three instants, 1/100, 2/100, and 3/100 of a second later? Again, please sketch carefully.

2. (*a*) Would you increase or decrease the frequency of the wave generator in a ripple tank in order to produce waves of greater wavelength?

(*b*) In a ripple tank, when one crest (and one trough) is sent every 1/10 of a second, the wavelength λ is 3 centimeters. What are the frequency and speed of the waves?

3. Sound waves in air usually travel about 340 meters/second. Sounds audible to the human ear have a frequency range from about 30 to about 15 000 oscillations/second. For such sound waves, what is the longest wavelength? The shortest?

Compare those wavelengths with distances in your dorm or on your desk. One doesn't usually think of this while talking, does one?

4. The speed of light as measured in a certain transparent solid is 2×10^8 meters/second. (The Newtonian particle theory is dead and irrelevant here.)

(*a*) Suppose a ray of light comes from vacuum onto the solid, as in sketch (*a*) of figure 3.19. Will the ray bend away from the normal

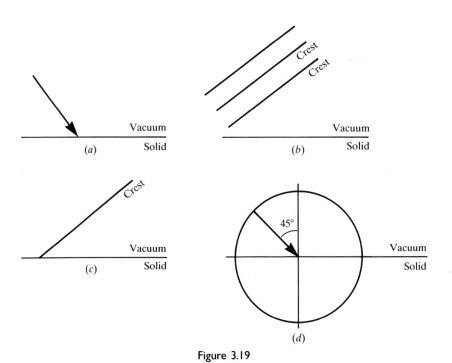

Figure 3.19

78

or toward it? Reproduce the sketch carefully and show the refracted ray.

(b) Sketch (b) shows several crest lines of a light wave at one instant of time. Reproduce the sketch, using a spacing of 1 centimeter between the crest lines. Then draw the crest lines at a time 10^{-11} seconds later.

(c) Sketch (c) shows a crest line of a light wave at one instant of time (as the light heads toward the solid). Reproduce the sketch and then draw the crest line at a time 10^{-10} seconds later. (Note that the time interval here is different from the interval in the previous part. You may ignore the reflected crest line.)

(d) Are your responses to parts (a) and (c) consistent? Explain briefly.

(e) For the situation in sketch (d), calculate either the semi-chord of the refracted ray or the sine of the angle of refraction. Indicate them on a careful reproduction of the sketch.

4   Interference

In making some experiments on the fringes of colors accompanying shadows, I have found so simple and so demonstrative a proof
of the general law of the interference of two portions of light,
which I have already endeavored to establish, that I think it right
to lay before the Royal Society a short statement of the facts which
appear to me so decisive.

Thomas Young,
Bakerian Lecture, November 24, 1803

4.1   Interference defined

In chapter 3, we framed the question, can we account for the behavior of
light by supposing that light is wave motion of some kind?

So far, the logic of our response has been of this form: we asked
experimentally, do waves have property X that we know light possesses?
For "property X," we took

(1) the existence of reflection,

(2) the equality of the angles of reflection and incidence,

(3) the existence of refraction, and

(4) Snell's law, in the sense that, in refraction, the ratio of sines is
independent of the angle of incidence.

Each time, we could answer, yes, waves do have this property X that we
know light possesses.

Beyond these tests, we found that the wave theory's prediction for the
index of refraction – namely, a specific ratio of wave speeds – was resound-
ingly confirmed by Michelson's measurements of v_{air}/v_{water}.

Things are going well for a wave theory of light. The time has come to
shift the perspective and to change the logic of the testing. Now we ask,
does light have property Y that we know waves possess?

For "property Y," we take the superposition principle: to get the
resultant pattern during overlap, add the original patterns algebraically.

A glance back at figures 3.4 and 3.5 reminds us of the essentials of the superposition principle. When two crests meet, they reinforce each other, and a larger crest arises. We will call this *constructive interference*.

When a crest and a trough meet, the algebraic addition of those two waves produces a zero wave (momentarily, anyway). The crest and trough cancel each other. We will call this *destructive interference*. The word "interference" seems quite appropriate here, for the crest and trough interfere with each other (in the colloquial sense) so much that they momentarily annihilate each other. One might think that physics would speak of "destructive interference" and "constructive collaboration," but the noun "interference" is used in both cases, the adjectives serving to distinguish the two behaviors.

How should we describe the meeting of two troughs? If you turn figure 3.4 around by 180 degrees and look from the new bottom to the new top, you will find the answer. When two troughs meet, they interfere constructively, producing a deeper trough.

In short, crest and crest give constructive interference, as do trough and trough. Only crest and trough cancel each other and give destructive interference.

These two kinds of interference – constructive and destructive – are almost synonymous with the superposition principle. Thus we can phrase the central question of this chapter as follows: does light have the property of interference that we know waves possess? An affirmative answer will greatly support the proposition that light is wave motion of some kind.

4.2 Two-source interference

So far in this book, our experience with interference comes from waves in one-dimensional motion, specifically, crests and troughs moving along a metal skeleton. Before we investigate light, we need to extend our understanding of interference to wave motion in two dimensions. A ripple tank provides a good way to do that.

If you dip your index finger periodically into the water, you form a sequence of concentric ripples; crest and trough follow each other regularly in expanding rings. If you use the index fingers of both hands, separated by a hand's width, then you form two sets of ripples. Where the ripples overlap, interference takes place.

A mechanical gadget will dip into the water more regularly than you can with your fingers. Figure 4.1(*a*) shows the periodic waves that are produced by two little spheres dipped regularly into the water of a ripple tank.

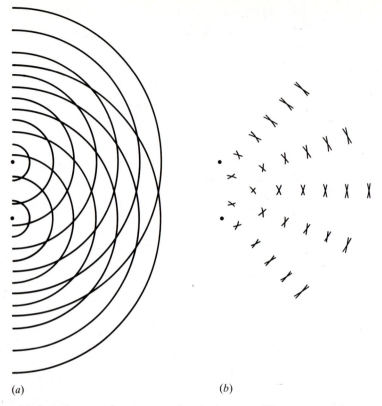

(a)   (b)

Figure 4.1 Periodic waves from two synchronized sources. The waves are shown at one instant of time, after seven crests have been sent forth. (a) The sequences of crest lines, one sequence from each source. (b) The locations where a crest line from one source overlaps a crest line from the other and produces constructive interference.

Actually, only the crest lines of the waves are shown, and even those only at one instant of time, after seven crests have been sent forth. (Otherwise the diagram would be hopelessly cluttered.) The intersections in part (a) tell us where a crest line from one source overlaps a crest line from the other. Those intersections produce constructive interference. Your eye can probably discern a pattern to those intersections. Part (b) displays the intersections in isolation; strings of them are evident.

A periodic wave has a trough between each pair of adjacent crests. To include the troughs and still retain clarity, we need to select a portion of the scene and magnify it. Figure 4.2 does that. Where a trough line from one source overlaps a crest line from the other, the waves interfere destructively. As part (b) of the figure shows, the locations where the waves cancel each other also form a pattern. Strings of destructive intersections lie between strings of constructive intersections.

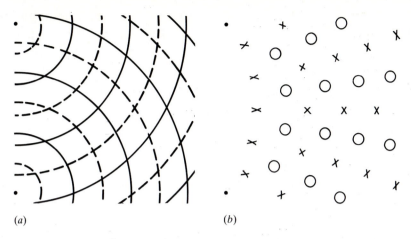

(a)   (b)

Figure 4.2 A magnified view of some crest lines and trough lines. (a) The trough lines are indicated by the dashed lines. (b) As before, the crosses denote locations of constructive interference. The open circles symbolize locations of destructive interference.

The overall pattern of intersections that we see in these snapshots persists in time. An actual experiment or a film loop provides the most persuasive evidence, but we can reason it out, too, at least for the central line, as follows.

The central line extends directly to the right from the point half-way between the two sources. (In more geometric terms, the central line is the perpendicular bisector of the short line joining the two sources.) Refer to figure 4.2(a), and let us focus our attention on the right-most point (along the central line) where two crest lines overlap. At the time of the snapshot, that point was a location of constructive interference. As time goes on, the ripples expand; the crest lines move outward, and soon two trough lines will overlap at that same point. That, too, is constructive interference: a deeper trough. The point in space on which we rivet our attention is equidistant from the two sources. So when a crest reaches the point from one source, a crest must reach it from the other. The same is true for a trough from one source and a trough from the other. And for portions of a wave that are intermediate between a crest's maximum and a trough's nadir – for example, a wave height of two-thirds the crest's maximum – again the arrival from one source will be matched by an arrival from the other. Except for those brief moments when the ripple is precisely between its crest half and its trough half and hence does not shift the surface of the water at all, the contributions from the two sources will add constructively. Thus constructive interference persists at this point (and at all other points along the central line).

The complexity of this section mandates a summary paragraph. When two sources send out periodic waves in synchronism, a persistent pattern of interference develops in space. All along the central line, there is constructive interference. To either side, lines of destructive interference alternate with lines of constructive interference. The locations of these lines remain fixed in space.

4.3 Interference in light?

Now we are ready to ask whether light has the property of interference. For the equivalent of two synchronized sources, we can permit light from a single source to fall on a pair of closely-spaced slits in an opaque card. If light is a wave of some kind, the light emerging from the two slits will act like waves from two synchronized sources. A red filter placed over a non-frosted straight-filament lamp provides a simple source of red light. A good size for the slits is 1 centimeter long and separated by 0.13 millimeters, that is, 0.13 millimeters from the center of one slit to the center of the other. The width of each slit might be one-third of their mutual center-to-center separation. Thus the slits are indeed narrow and closely spaced; figure 4.3(a) provides a sketch. A satisfactory distance between the filament light source and the slits can be as short as 50 centimeters or as long as 15 meters (if the source is bright).

If you hold the slits a centimeter from one eye (so close that the card touches your forehead), you see a pattern of bright red lines, as illustrated

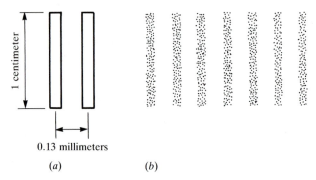

1 centimeter

0.13 millimeters

(a)   (b)

Figure 4.3 Double-slit interference in red light. (a) The dimensions for a good pair of slits. In the drawing, the vertical and horizontal scales differ. (b) The pattern of bright and dark on your retina. Stippling represents brightness. A similar pattern appears on a distant wall when the slits are illuminated by a laser beam. In both situations, more brights than the seven shown here may be visible.

in figure 4.3(*b*). The most striking feature is this: although there are only two slits, more than two bright lines appear.

The multiplicity of bright lines does not arise from some peculiarity of our eyes. If the slits are clamped vertically on a table and illuminated with the circular red beam from a neon laser, a pattern of bright spots appears on a distant wall, many more than two spots.

Both on your retina and on the wall, you can discern a brightest central line (or spot) and then other lines (or spots), less bright, arrayed symmetrically on each side.

If we suppose – tentatively – that red light is a periodic wave of some sort and then look back at figures 4.1 and 4.2, we can make sense of the many bright lines (or spots). Where the waves (now imagined to be coming from two slits) interfere constructively, the light should be intense, thus producing the bright lines (or spots). And where the waves interfere destructively, there should be no light, thus accounting for the intervening dark regions.

Figure 4.4 provides a top view of the situation. A periodic sequence of crests comes in from the left; this is a tentative description of the laser beam. Adjacent crests are separated by the wavelength λ. The opaque card appears edge-on. The slits extend above and below the plane of the page; the gaps in the card represent the tiny widths of the slits. From each slit will emerge circular crest lines, just as in figure 4.1. The point P is a general point on the wall where the pattern of bright spots appears. We can move the point P around – mentally – and ask, at which locations for P does

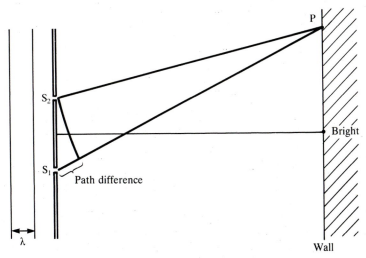

Figure 4.4 Double-slit interference: the path difference. The symbols S_1 and S_2 denote the locations of the two slits.

constructive interference occur? And at which locations does destructive interference occur?

Where the central line meets the wall, constructive interference will occur. After all, that junction is equidistant from the two slits, and so the arrival of a crest from one slit will coincide with the arrival of a crest from the other. Thus that location is labeled "Bright."

Next, let us return momentarily to figure 4.1 and inspect it carefully. Where the crest lines overlap, constructive interference occurs. One crest line comes from each source. Except along the central line, the distances from an overlap point to the two sources differ. The distances differ by one wavelength or by two. For example, we can find the point where the second crest line emitted by the upper source overlaps the first crest line emitted by the lower source. The difference in distances is one wavelength. We may generalize: so long as the difference in distances is an integral number of wavelengths, constructive interference will occur.

For destructive interference, we return to figure 4.2. Where a trough line from one source overlaps a crest line from the other, the distances from the sources differ by $\frac{1}{2}$ wavelength or $1\frac{1}{2}$ wavelengths.

Now we return to figure 4.4. As sketched, the point P is farther from the location S_1 than from S_2. The difference in distances is $S_1P - S_2P$; we will call that difference the *path difference*.

The path difference determines whether the point P will be a point of brightness or of darkness.

It will be bright at point P if a crest from S_2 arrives at P at the same time as a crest from S_1. Crests are separated by a distance λ. Hence we can say

bright at P when the path difference equals 0, λ, 2λ, 3λ, ..., that is, any integer times λ.

It will be dark at P if a crest from S_2 arrives at P at the same time as a trough from S_1. A trough lies half-way between adjacent crests. Hence we can say

dark at P when the path difference equals $\frac{1}{2}\lambda$, $1\frac{1}{2}\lambda$, $2\frac{1}{2}\lambda$, ...

The path difference really matters where the two paths come together. Out there is where a trough from one slit may overlap a crest from the other, giving darkness. Or crests from both slits may overlap, perhaps the second crest to emerge from S_2 with the first crest from S_1, giving brightness. The path difference is most easily shown, however, at the slits' end of the two lines. If we draw an arc with P as center and radius S_2P, the arc cuts the line S_1P and shows us the extra length of the longer line; that is the path difference. If it equals λ, then one extra wavelength fits in, and the point P will exhibit crest on crest and hence brightness. If the path difference equals

$\frac{1}{2}\lambda$, then one-half of an extra wavelength fits in, and so the arrival at P of a crest from one slit must be accompanied by the arrival of a trough from the other slit. Thus a path difference of $\frac{1}{2}\lambda$ produces destructive interference and darkness. And so on for other specific values of the path difference, as noted earlier.

There is some complexity in the many possibilities for constructive and destructive interference. For the purposes of the next few pages, one item will suffice. Here it is: the first off-center bright occurs where the path difference is one wavelength, that is, $S_1P-S_2P=\lambda$. And the reason is this: another complete wavelength can fit along the longer of the two paths, and so again there is constructive interference.

A quick experimental test

It looks as though the idea of wave interference can explain the red lines (or spots). Moreover, we can quickly subject this tentative explanation to a test. Using the interference idea, we predict what should happen to the pattern if we increase the separation between the slits (by a factor of 3). And then we see what experiment says.

Figure 4.5 shows us what we need to know. At the central bright, the path difference from a pair of slits is zero. For the first off-center bright, the path difference must be one wavelength. After the slit separation has been increased, a path difference of one wavelength arises closer to the central bright. Thus we predict that the interference pattern shrinks.

If you hold the original slits and then the more widely-spaced slits to your eye, you see the pattern shrink – as predicted.

We find that light has the "property Y" of interference that we know waves possess. We can be confident that light has a wave-like character.

Before we go on, a name is in order. The interference that we have studied in this section – interference by light emerging from two slits – is called *double-slit interference*.

4.4 Measuring the wavelength of light

In double-slit interference, a path difference of one wavelength produces the first off-center bright spot. This fact should enable us to measure the wavelength of light. A direct route would be to measure the length S_1P, measure the length S_2P, and subtract. That, however, is a hopeless proposition. Here is the reason why. Already the separation of the slits is small. A path difference of one wavelength is much less than the slit separation, and

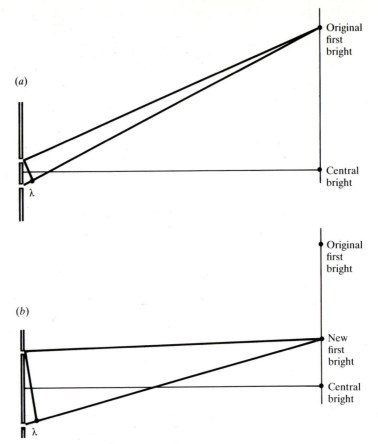

Figure 4.5 How a change in slit separation affects the location of the first off-center bright. Both path differences are one wavelength of red light. (a) Original slit separation. (b) Slit separation three times as large. If lines were now drawn to the location of the original first bright, the path difference would be larger than one wavelength.

so is extremely short. The direct route would be like trying to measure the length of a paperclip lying on your desk by measuring the distance from each end to a certain stone in the Washington Monument and then subtracting. It is not practical.

Rather, we need to ask, "in this context, what *can* we measure?"

Figure 4.6 suggests some possibilities. We can readily measure the distance L along the central line from the slits to the wall, and we can measure the distance $y_{1st\,bright}$ along the wall from the central bright to the first off-center bright. The separation of the two slits can be measured with a microscope. These three values – plus some geometry – suffice, as follows.

First, we establish a pair of similar right triangles. One right triangle is formed by the length L, the length $y_{1st\,bright}$, and the diagonal to P from the

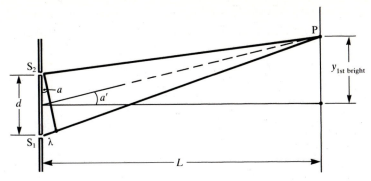

Figure 4.6 The geometry for measuring the wavelength of light. If a neon laser provides the light, then realistic values are $L=10$ meters, $d=0.13$ millimeters, and $y_{1st\,bright}=4.8$ centimeters. When L is much larger than d (as would then be true), the three diagonal lines are almost parallel.

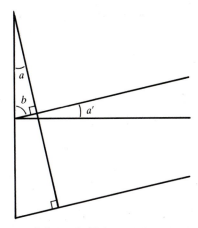

Figure 4.7 An enlargement to help in establishing similar triangles. The angle marked b is a useful auxiliary angle.

point mid-way between the slits. The other triangle is formed by the separation d, the path difference λ, and the arc swung from S_2 over to the diagonal S_1P. When L is much greater than d, the arc is almost a straight line, yielding a right triangle. To prove the triangles similar, we need only prove that angle a' equals angle a.

Figure 4.7 shows the crucial area magnified. The angles a' and b add up to 90 degrees. Because angles a and b share a triangle with a right angle, the sum of angles a and b must also be 90 degrees. Thus angles a' and a are equal. We return to figure 4.6. The equality of angles a and a' in triangles that are right triangles proves that the triangles are similar. (That is, the triangles have exactly the same shape, one being just a large-scale version of the other.)

Second, for each triangle, we form the ratio of the short side to the hypotenuse. The similarity of the triangles implies that the ratios are equal:

$$\frac{\lambda}{d} = \frac{y_{1st\ bright}}{\sqrt{L^2 + y_{1st\ bright}^2}} \approx \frac{y_{1st\ bright}}{L}. \tag{1}$$

For the hypotenuse of the larger triangle, we use the Pythagorean theorem. In practice, the length $y_{1st\ bright}$ is much smaller than L; so we may drop $y_{1st\ bright}$ under the square root sign, and then a tidy result emerges. The approximation is so good that we will subsequently regard it as an equality.

Multiplication by d on both sides yields the relation

$$\lambda = \frac{y_{1st\ bright}\ d}{L}.$$

As promised, measuring the three quantities on the right-hand side does suffice to determine the wavelength.

When all lengths are expressed in meters, the values quoted in the legend for figure 4.6 yield

$$\lambda = \frac{4.8 \times 10^{-2} \times 1.3 \times 10^{-4}}{10}$$

$$= 6.2 \times 10^{-7}\ \text{meters.}$$

This value, typical of what a lecture demonstration produces, is a remarkably small length.

To grasp how small the wavelength is, let us compare it with the thickness of a thumbnail. The nail might be half a millimeter thick: 0.5×10^{-3} meters. How many wavelengths could one fit along the thickness of a thumbnail? We divide the thickness by λ and find

$$\frac{\text{Thickness}}{\lambda} = \frac{5 \times 10^{-4}}{6.2 \times 10^{-7}} \approx 10^3.$$

For a rough estimate, we may cancel the 5 with the 6.2; then we multiply numerator and denominator by 10^{+7} and combine exponents. Thus about 1000 wavelengths of red light would fit along the thumbnail's thickness. The wavelength is tiny indeed, and so it is no wonder that we need to measure it indirectly.

Two remarks are in order before we move on. First, when the red light from a neon laser is measured with more accurate apparatus, the wavelength emerges as 6.328×10^{-7} meters. The double-slit method we discussed determines the exponent and the first significant digit correctly – and that is sufficient for our purposes.

Second, equation (1) of this section can be generalized to describe the location of the second, third, and other brights in terms of L, d, and λ. The path difference that generates the nth off-center bright is $n\lambda$. Thus the short side of one triangle changes from λ to $n\lambda$, and the short side of the other, from $y_{\text{1st bright}}$ to $y_{n\text{th bright}}$. The new ratios imply the relation

$$\frac{n\lambda}{d} = \frac{y_{n\text{th bright}}}{L},$$

to ample accuracy. The location of the nth bright is proportional to n, and so the brights are equally spaced along the wall.

4.5 Color and wavelength

Interference enables us to study the relationship of color and wavelength. For example, does blue light have a wavelength shorter than that of red light or longer or the same?

Looking through slits at blue and red light simultaneously will provide the comparison. A vertical straight-filament lamp with a red filter over the top half and a blue filter over the bottom half provides the two colored lights. If you look with the double slits that were described in the preceding sections, no clear distinction between red and blue interference patterns is evident. Things are too cramped, and so we need to spread the pattern out and make it sharper.

Making the slit separation d smaller spreads the pattern out. And using many slits, rather than just two, makes the pattern sharper. Eighty slits, separated center-to-center by 3×10^{-5} meters, provide a clear distinction on your retina: the off-center brights in blue light fall closer to the central bright than do the off-center brights in red light.

Figure 4.8 shows what would appear on a wall if such slits were illuminated by two laser beams, one blue and the other red. For the two slits shown there, the paths to the central bright would be equal. As one mentally shifts attention outward along the wall, the path difference grows. The image that you had on your retina tells us that a path difference equal to λ_{blue} arises sooner than a path difference equal to λ_{red}. Therefore λ_{blue} is less than λ_{red}.

In general, short wavelengths produce constructive interference closer to the central bright than do long wavelengths.

Table 4.1. *Wavelength ranges and the associated colors. Various authors and authorities place the boundaries slightly differently, and so no table is definitive. Red may go a bit longer than 7.0×10⁻⁷ meters, and violet a bit shorter than 4.2×10⁻⁷ meters. The boundary between red and orange is sometimes placed at 6.2×10⁻⁷ meters*

Color	Wavelength range (in meters)
Red	$(6.1\text{--}7.0)\times10^{-7}$
Orange	$(5.9\text{--}6.1)\times10^{-7}$
Yellow	$(5.7\text{--}5.9)\times10^{-7}$
Green	$(5.0\text{--}5.7)\times10^{-7}$
Blue	$(4.5\text{--}5.0)\times10^{-7}$
Violet	$(4.2\text{--}4.5)\times10^{-7}$

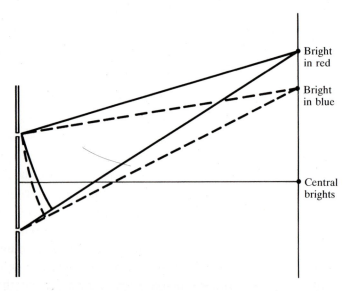

Figure 4.8 Comparing the wavelengths of blue and red light. For clarity's sake, only the central two of many slits are shown, and only the first off-center brights in each color. Imagine that blue and red laser beams illuminate the slits. Recall that the first off-center bright arises where the path difference equals one wavelength.

If the filters are removed from the filament lamp, your view through the multiple slits gives the rainbow colors of white light, a full spectrum for each off-center bright. This time the colors are spread out according to wavelength. Table 4.1 provides detail on wavelength ranges and the associated colors.

Each color in the rainbow corresponds to a specific wavelength. (Because these colors appear in the spectrum of a glass prism or of multiple slits, they are called the *spectral* colors.) To be sure, there are many different reds in the "red" wavelength range, 6.1×10^{-7} to 7.0×10^{-7} meters. They vary subtly in hue. The "red" light from a neon laser, having a wavelength of 6.328×10^{-7} meters, is just one of many spectral reds. The point of the paragraph is this: if we isolate by eye a specific red from the red portion of the rainbow, then that red is produced by light of a single wavelength (or by light having a quite narrow band of contiguous wavelengths). The same is true for an orange or yellow or green or blue or violet from the rainbow.

Beware, however. Not every color that the eye perceives is associated with some single wavelength. For example, purple is always a mixture of light from the red end of the spectrum with light from the blue and violet end. Some yellows correspond to single wavelengths, but a thoroughly creditable yellow can be produced by mixing red and green light, for example, light with $\lambda_{red} = 6.5 \times 10^{-7}$ meters and $\lambda_{green} = 5.2 \times 10^{-7}$ meters. Human color vision arises from a complex interplay of physics, physiology, and psychology.

4.6 Thomas Young

This section picks up the historical thread where we dropped it after learning about Newton's achievements and his particle theory of light.

A rival wave theory existed in Newton's day. Its foremost advocate was Christian Huygens, an especially thoughtful Dutch physicist. His theory, however, was seriously incomplete; it dealt with only a small fraction of the phenomena that Newton encompassed in his *Opticks*. Moreover, the theory was difficult to understand in its details, and it certaily contained no idea of periodic waves. It was as though an oceanographer thought only of tidal waves – a single huge crest – and was oblivious to the everyday periodic waves that wash up many times a minute on beaches all over the world. No wonder that wave theories of light were held in low regard during the 1700s, while Newton's particle theory held center stage.

To be sure, the adherents of Newton's theory were unable to make progress on the few problems that Newton had bequeathed them as unsolved puzzles, but never mind. The wave theory had nothing compelling to offer in opposition – nothing, that is, until Thomas Young discovered interference in 1801.

Neither Newton nor Einstein was a child prodigy. Their inherent genius began to show itself only when they were in their twenties. Thomas Young was unquestionably different. At the age of two, he read English easily. A doting grandfather plied him with books, and, four years later, the precocious Thomas started Latin. After all, he had finished his second complete reading of the Bible in English. By the age of 16, he knew Latin, Greek, French, Italian, Hebrew, Chaldean, Syriac, Samaritan, Arabic, Persian, Turkish, and Ethiopic. Archeology was obviously one of his interests.

At age 19, Young decided to enter medicine. He studied in London, Edinburgh, and Göttingen, receiving his M.D. from the German university at age 23. He practiced medicine in London – but never very successfully, for his interests lay in languages, human vision, and physics.

By age 21, Young had been elected a member – a "fellow" – of the Royal Society. What brought his election at so early an age? He had discovered how the eye focusses on objects at different distances, for example, on the page you are reading versus a distant mountain top. Young explained that muscles within the eye lens contract or relax, pulling the lens into a more strongly convex shape or allowing it to become flatter. Such change in shape alters the amount by which the lens refracts light and so alters the image that it produces. For an object at a specific distance, there is a specific amount of refracting that will place the image sharply on the retina.

In 1804, Young was appointed Foreign Secretary of the Society. Surely he had no trouble corresponding with scientists on the Continent.

You may recall the Rosetta stone, found in Egypt by the French in 1799. Carved on it is the same message in three languages: Greek, Egyptian demotic (the simplified, common script, used for books and deeds), and Egyptian hieroglyphics (perhaps best remembered for the avian symbols – falcon and ibis – that appear so often). The message itself is a decree of the priests assembled at Memphis on the Nile. Young contributed substantially to the decipherment of the demotic script. (Just how much is a matter of dispute among the experts.)

Thomas Young had a great admiration for Newton and knew his *Opticks* thoroughly. As he puzzled over some of the experiments Newton described, Young realized that he could make complete sense of them if he assumed light to be some kind of periodic wave motion and if he added a crucial ingredient; as he said in 1801,

> When two Undulations . . . coincide either perfectly or very nearly
> in Direction, their joint effect is a Combination of the Motions
> belonging to each.

In its content, this statement is what we called the superposition principle for waves; we expressed the same idea in the words, "to get the resultant pattern during overlap, add the original patterns algebraically." And we noted that from the superposition principle follow the phenomena of constructive and destructive interference. Interpreting some of Newton's data in this framework, Young concluded that the wavelength of red light is 6.5×10^{-7} meters and that the wavelength of violet is 4.4×10^{-7} meters, both in good agreement with the values recognized today. Indeed, the good correspondence speaks well for both Young's theoretical insight and Newton's experimental acumen.

Young used his idea of wave interference to investigate light in a number of contexts; some of them will appear in section 4.8. A few years after his great insight, he devised the experiment for which he is most famous: double-slit interference. Apparently Young did the experiment both with two pinholes in place of two slits and also with two slits, as we have discussed in this chapter. In each case, as he wrote in 1807, the light on a viewing surface "is divided by dark stripes." He goes on:

> The middle . . . is always light, and the bright stripes on each side are at such distances, that the light coming to them from one of the apertures must have passed through a longer space than that which comes from the other by an interval which is equal to the breadth of one, two, three, or more of the supposed undulations [which we would call "wavelengths"], while the intervening dark spaces correspond to a difference of half a supposed undulation, of one and a half, of two and a half, or more.

The explanation is succinct and correct.

It would be nice to report that the world of physics recognized the error of its ways and converted from a particle theory of light to a wave theory. Such was not the case. Newton's particle theory held too strong a grip on the imagination. It was not until Augustin Fresnel, a French civil engineer, independently developed a wave theory and its associated mathematics in the years 1815–20 that the resistance began to crumble. An anecdote about Fresnel concludes this chapter, but, before that, the physics of a single slit merits our attention.

4.7 The single slit

By now we are familiar with the interference pattern produced when light passes through two slits. Yet light, it seems, is always ready with another surprise.

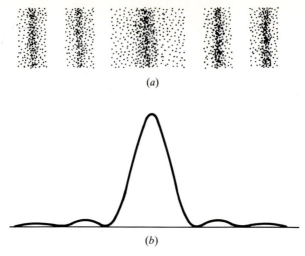

(a)

(b)

Figure 4.9 The pattern produced by red light going through a single narrow slit. (a) As the eye perceives the pattern. (b) What an instrument measuring intensity would record. Intensity is plotted vertically, and location, horizontally.

We return to the straight-filament lamp, covered again with a red filter. If you view the lamp through only a single narrow slit in an opaque card, again a pattern of bright lines appears. Figure 4.9 provides a sketch of how the light intensity varies across your retina. There is a broad central bright and then subsidiary brights arrayed symmetrically to the sides. One slit, but many brights. Careful comparison with a double-slit pattern reveals a difference. Here the central bright is about twice as broad as the adjacent off-center brights; in the double-slit situation, all brights have about the same width. The two patterns show a family resemblance, however, and so interference is probably the key to the single-slit pattern, too.

If we clamp a card with a single slit to a table and illuminate it with the beam from a neon laser, the spots on a distant wall exhibit the same pattern that figure 4.9 shows for your retina. The situation with laser, slit, and wall is easier to analyze geometrically, and so we concentrate on it.

A central bright spot is easy to accept. The challenge is to figure out why off-center brights should arise. We meet the challenge in two steps.

First, we use interference ideas to understand why the first dark (adjacent to the central bright) ought to arise. Figure 4.10 sets the scene. Point P is chosen so that a new path difference will be exactly $\frac{1}{2}\lambda$. The path difference compares the distance "mid-point-of-slit to P" with the distance "edge-of-slit to P." Mentally, we divide the slit's width into eight tiny intervals; they are numbered in the sketch.

Crest lines and trough lines of light come through the slit from the left

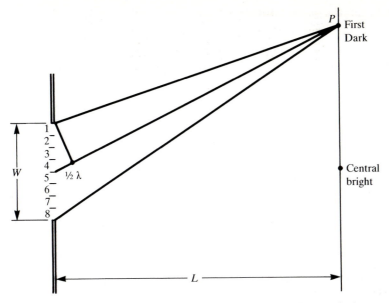

Figure 4.10 The analysis for the single-slit pattern. The slit's length extends above and below the plane of the paper. In reality, the distance L is very much larger than the slit width W. Thus the three diagonal lines are almost parallel.

and spread out into the region of the drawing. Spreading is characteristic of waves. (Recall the expanding ripples that arise when you toss a pebble into a mud puddle.) As the crest lines and trough lines spread out, they reach point P. Focus attention on intervals 1 and 5. Their distances from point P differ by $\frac{1}{2}\lambda$. When a crest line from interval 1 reaches P, a trough line from interval 5 also reaches P; the two interfere destructively, canceling each other. Now we shift our attention to intervals 2 and 6. Because the diagonal lines are practically parallel, the path difference for intervals 2 and 6 is also $\frac{1}{2}\lambda$. Thus, whatever effects the intervals 2 and 6 produce at P, the effects cancel each other. Similarly, we can pair for cancellation the intervals 3 and 7 and then the intervals 4 and 8.

In short, point P is a location of complete destructive interference for waves coming through the single slit. We were able to establish this result because we paired the eight intervals judiciously and found then to cancel by pairs. If we tried a different pairing, for example, interval 1 with 2, then 3 with 4, and so on, we would not be able to figure out what happens at point P. That is all. What actually occurs at P does not depend on how we mentally divide up the slit's width and then pair intervals, but only one method provides a direct route to the answer. (To be sure, dividing the slit into a different even number of intervals works also. The larger the number of intervals, the better, for then variations within an interval are sure to be unimportant.)

Complete destructive interference requires a carefully-orchestrated mutual cancellation of all contributions. It is special. We have found one occurrence: when the path difference computed from the lengths "edge-of-slit to P" and "mid-point-of-slit to P" is one-half wavelength.

(A look at figure 4.10 gives us an equivalent formulation. The path difference computed from *both edges* of the slit to point P is double what we have described for the path difference from one edge and the mid-point. Hence the first dark occurs when the path difference computed from *both edges* of the slit to point P is one entire wavelength.)

But now for the second step. If we mentally move point P a bit farther from the central bright, we move away from a special location. We may be confident that complete destructive interference no longer occurs, and some degree of brightness arises on the wall.

Still farther along the wall there may be other special locations that produce complete destructive interference; between them would be regions of brightness. That is indeed the case, but we need not work it out. Suffice it to say that interference is quite capable of explaining why light passing through a single slit produces many brights.

Do it yourself

You do not need any special apparatus to see single-slit interference. A street light or even a car headlight at 100 meters can serve as a light source. Your index and middle fingers become the opaque card, the gap between them being the single slit. With your palm toward your face, hold the two fingers a centimeter or so in front of one eye and look between the fingers at the light; close your other eye. The only trick is to adjust the slit width. To do that, grip the tops of the fingers with your other hand and slowly squeeze the two fingers together. Start with a gap wide enough so that you can see the light and some surrounding terrain. Just when you fear that the gap will vanish and nothing will come of your attempt, some off-center brights will appear. Most likely, they will be wiggly lines on both sides of the bright center. Shifting your fingers up or down a little may improve the image. As you reduce the gap further – but slowly – the central bright will spread out. By alternately squeezing and releasing your two fingers, you can make the pattern pulsate.

I am embarrassed to confess how often I have stopped along a street, struck by a particularly good light source, and checked that light still shows interference and that it is indeed a wave phenomenon.

4.8 More interference phenomena

The interference of light is not rare in our lives. Rather, we are often unaware that interference is the explanation of some common observations.

Take the soap bubbles that children blow and that you may have blown. You dip a circular loop into a soap solution and wave it in the air. Magically, several bubbles appear and drift lazily through the summer air. As the sunshine reflects from the bubbles, they shimmer with pale green and pink hues.

In greatly exaggerated fashion, figure 4.11 shows what is happening. When light from the sun strikes the bubble hemisphere facing you, some light is immediately reflected to your eye (by the outer surface of the soap film). Another portion of the light is refracted into the soap solution, travels the tiny thickness of the soap film, and is reflected at the inner surface of the film. That reflected portion returns to the outer surface, and some of it is refracted out of the film and travels to your eye. On your retina, the two beams can interfere, constructively or destructively. For a certain thickness of soap film and a certain angle of travel through the film, the green portion of the sun's spectrum will enjoy constructive interference; the red portion,

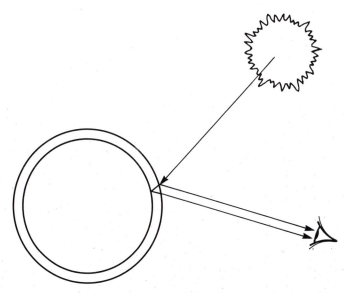

Figure 4.11 How interference arises when sunlight reflects from a soap bubble (drifting in mid-air). For clarity's sake, the thickness of the soap film is much exaggerated, and only the essential light paths are displayed. In reality, the soap film is extremely thin relative to the bubble's radius, and so the two rays reflected to the eye run nearly parallel to each other and very close together.

having a significantly different wavelength, will annihilate itself by destructive interference; and the corresponding region of the bubble will appear greenish. (The other colors in the spectrum do something intermediate between full constructive interference and destructive interference; often they just provide a pale luminous background.) For a different thickness or angle, the red section of the spectrum will march into your eye, crest on crest, along the two paths, and so the corresponding region of the bubble will appear pink.

To be sure, green and pink are not the only colors that ever appear. Sometimes the bubble presents a study in peacock blue and citron yellow. The explanation is still interference, but different portions of the spectrum interfere constructively.

The very same explanation serves for the iridescent oil slicks you see on a road or driveway after a thundershower. The rings are often 20 centimeters or so in diameter, though they can be larger. The periphery often consists of pink or greenish rings. Beams of sunlight, reflecting from both the top and the bottom of the thin oil film, interfere on your retina. The thickness of the film and the angle determine which portions of the spectrum have the correct wavelength to interfere constructively and hence dominate the appearance.

The astrophysicist Marcel Minnaert was fascinated with color in the world around us. His classic little volume, *The Nature of Light and Colour in the Open Air*, is a goldmine of information and anecdotes. After giving some details about oil spots and their colors, Minnaert has this to say:

> We shall not be quite satisfied until we have made coloured rings ourselves. A drop of paraffin [kerosene] or a drop of turpentine poured on to a pond produces indescribably lovely colours! But if we use *oil* for this experiment, we shall find a surprise in store for us! The oil does not spread out into a film, and we see nothing. The same thing happens on a wet road as on a surface of water. Are the spots in the road due to petrol [gasoline] perhaps, not to oil? Again we are disappointed, for petrol produces only greyish white spots, apparently extremely thin and bearing no resemblance whatever to the magnificent rings of colour. Closer investigation has shown that only the *used, oxidised oil* dripping from the engine is capable of spreading on a wet surface. The more complete the oxidation of the oil, the thinner the layer becomes.

The iridescence of oil spots is not so easy to explain after all.

If you will turn back to figures 1.16 and 1.17 for a moment, they will remind you of the striking bright and dark rings produced by a thin sheet of mica. In figure 1.16 we find the rays that interfere and produce the patterns. Two rays emerge from the mica, heading rightward, and interfere to produce the pattern in transmitted light. Two other rays are reflected leftward and make – by interference – the pattern in reflected light. To be sure, many rays emerge from the sodium light source, and a detailed computation is difficult, but the drawing gives us the central idea: the bright and dark rings arise by interference of light.

A soap film, an oil slick, and a sheet of mica: each provides a thin layer of something, and light will reflect from both surfaces of the layer. The two reflected rays – one from each surface – will interfere, either constructively or destructively or something in between, and so they will give rise to variations in brightness and color on a viewing surface or on your retina. Generically, the topic of these paragraphs is "interference produced by thin films."

The Poisson spot

The chapter closes with a good story, one of the best in the history of optics.

Two sections back, we learned that Thomas Young introduced and advocated the idea of interference in the first few years of the 1800s. Independently, Augustin Fresnel developed the idea – and extended it – a dozen years later. By 1817, the French Academy of Sciences felt the time was ripe for a definitive study of the phenomena encompassed loosely by Young and Fresnel. The Academy's method was to offer a prize for the best paper detailing an experimental study and advancing a theory that was supported by the data.

The competition was open to all of Europe, including Great Britain, but only two entries were submitted. The subject was precisely what Fresnel had spent the preceding few years in studying; he collected his data and mathematical theory into a stunning, comprehensive treatise of some 135 pages. The other entry came from a "nobody," and posterity is oblivious even of his name.

You will not be surprised to learn that Fresnel won the prize. The twist comes in how the committee of judges examined his essay.

Fresnel's theory led to mathematics that was difficult by the standards of his day. Sometimes Fresnel could set up the mathematics but not complete the solution. The judging committee, however, counted among its members the eminent mathematician Siméon Poisson. He was a staunch adherent of the Newtonian particle theory, and nothing would have delighted him more than to show Fresnel's wave theory to be untenable. Indeed, from Fresnel's theory, Poisson derived the following prediction:

Let parallel light impinge on an opaque disk, the surroundings being perfectly transparent. The disk casts a shadow – of course – but the very center of that shadow will be bright. Succinctly, there is no darkness anywhere along the central perpendicular behind an opaque disk (except immediately behind the disk). Indeed, the intensity grows continuously from zero right behind the thin disk. At a distance behind the disk equal to the disk's diameter, the intensity is already 80 per cent of what the intensity would be if the disk were absent. Thereafter, the intensity grows more slowly, approaching 100 per cent of what it would be if the disk were not present.

Poisson's prediction is obviously preposterous. This will surely destroy the wave theory – or so he must have hoped.

The chairman of the judging commission, François Arago, set up the experiment – and found the central spot of light. Figure 4.12 shows a modern version, done with laser light and with a steel ball ($\frac{3}{8}$ of an inch in

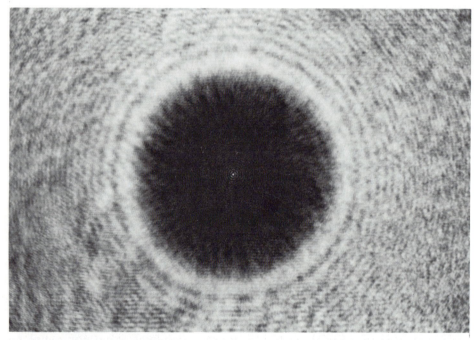

Figure 4.12 The "Poisson spot" is the central bright spot in the shadow cast by an opaque disk or sphere. The rings are more interference phenomena, also explicable by Fresnel's mathematics of interference. Just as light passing through a *single* slit produces brights and darks, so the light that passes by the opaque sphere makes concentric rings of bright and dark. The shadow, central spot, and rings show up nicely on a sheet of white paper taped to a wall. In this photo, the Poisson spot and interference rings were recorded directly on film (held in a 35-millimeter camera from which the lens had been removed).

diameter) in place of an opaque disk. As Arago reported to the French Academy of Sciences,

> One of your commissioners, M. Poisson, had deduced from the integrals reported by the author [Fresnel] the singular result that the centre of the shadow of an opaque circular screen must, when the rays penetrate there at incidences which are only a little oblique, be just as illuminated as if the screen did not exist. The consequence has been submitted to the test of a direct experiment, and observation has perfectly confirmed the calculation.

Additional resources

Marcel Minnaert's classic, *The Nature of Light and Colour in the Open Air*, is available from Dover Publications, New York. A much more recent book on the same subject – and equally delightful – is Robert Greenler's *Rainbows, Haloes, and Glories* (Cambridge University Press, New York, 1980).

John Worrall describes the history of the Poisson spot in "Fresnel, Poisson and the White Spot: the Role of Successful Predictions in the Acceptance of Scientific Theories," a chapter in *The Uses of Experiment*, edited by David Gooding, Trevor Pinch, and Simon Schaffer (Cambridge University Press, New York, 1989). John Strong notes that "the predicted spot had been discovered and announced by Maraldi over a half century earlier" but then had become lost to science; the reference is page 186 of Strong's *Concepts of Classical Optics* (W. H. Freeman, San Francisco, 1958).

An inexpensive but effective mechanical model for illustrating double-slit interference is described by your author and Vacek Miglus in the *American Journal of Physics*, volume 59, page 857 (1991). Seville Chapman, Director of the Physics Division, Cornell Aeronautical Laboratory, devised an excellent set of slits. I lend them to students in class several times a semester. Presently, the slits (protectively encased in glass) are available from Sargent-Welch, catalog number 3800A.

Questions

1. Figure 4.13 shows a double-slit interference experiment. The regions of darkness and (maximum) brightness in *blue* light are shown. Reproduce the sketch and then indicate where the regions of darkness and (maximum) brightness in *red* light would be.

2. A source of red light produces interference through two narrow slits spaced a distance $d=0.01$ centimeters apart. At what distance from the slits should we place a screen so that the first few interference brights are spaced

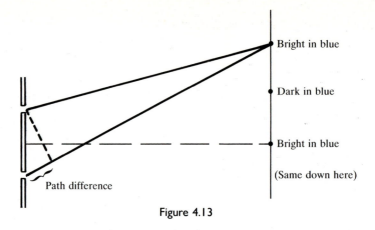

Figure 4.13

1 centimeter apart? Be careful to use a consistent set of length units: all lengths in meters or all in centimeters.

What will be the spacing of the brights if we then use violet light (with the screen still at the location used for red light)? To provide a qualitative check of your arithmetic, think about whether violet has a smaller wavelength than red or a larger wavelength and then think about whether the interference pattern should shrink or expand.

3. Suppose we do a double-slit interference experiment in a vertical tube, the slits at the top and the viewing surface at the bottom, as sketched in figure 4.14. With red light from our laser, say, we get a nice pattern of dark and bright stripes.

What would happen to the pattern if we *filled* the tube with water? Explain your reasoning. (Note: refraction is not relevant here.)

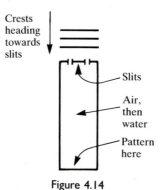

Figure 4.14

4. The visible light from a mercury vapor lamp passes through two slits whose separation is 0.002 centimeters. On a wall at a distance of 2 meters from the slits, first off-center brights in three colors are clearly visible. Here is a tabulation:

Color	$y_{1st\,bright}$ (in centimeters)
#1	5.79
#2	5.46
#3	4.36

(a) Convert *all* distances to meters (so that you will be assured of using a consistent set of lengths).

(b) Compute the wavelengths of the three colors.

(c) Use the table of wavelength interval versus color in this chapter to name the three colors.

5. To make a long, narrow slit, you can paint a piece of glass black and then scratch it once with a razor blade (held against a ruler). How could you use light itself to measure the width of the microscopically narrow slit that you made? (You do *not* have access to a microscope.)

6. A set of slits spreads out white light according to the wavelength. A glass prism, however, spreads out light according to its speed in glass (relative to the speed in vacuum). Each process spreads out the colors in the same rainbow order: red, orange, . . ., violet. (Or in the reverse of rainbow order. It depends on where you start looking.) What can you infer from this sameness of order?

7. Figure 4.15 shows the crest of a light wave at one instant of time and then the same crest somewhat later.

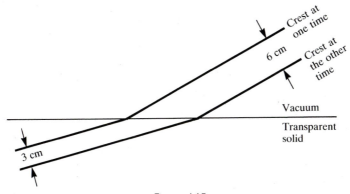

Figure 4.15

(a) What is the speed of light in the transparent solid?

(b) On the transition *from solid to vacuum*, would a light beam be bent toward the normal or away from it? On a reproduction of the sketch, show the normal and the beam's direction, both before and after the transition.

5 Electromagnetic waves

I have also a paper afloat, with an electromagnetic theory of light,
which, till I am convinced to the contrary, I hold to be great guns.

James Clerk Maxwell, in a letter
to his friend, C. H. Cay, January 5, 1865.

5.1 The electric field

Let us recall the plan we adopted at the close of section 3.1:

(*a*) Learn about waves.

(*b*) Find that light does indeed have a wave-like character.

(*c*) Learn what the "something" is out of which light waves are formed.

In subsequent sections, we found that waves do have the properties of
reflection and refraction that light possesses. Then we shifted perspective
and found that light has the property of interference that waves possess. In
short, chapters 3 and 4 taught us that light does indeed have a wave-like
character.

Before we can address part (*c*) of the plan, we need to know certain
aspects of electricity and magnetism. So we turn now to those twin subjects.

Sometimes, when you pull a sweater off over your head on a day when
the air is particularly dry (perhaps indoors in winter), you hear a crackling
sound and may even feel an electric spark. The frictional rubbing of the
sweater on your shirt or blouse separates some electric charges, one from
another; the crackling and the spark are a miniature form of thunder and
lightning.

More deliberate rubbing of two different objects separates electric
charges more effectively. Specifically, if one rubs a glass rod with an old silk
handkerchief, electrons from the rod remain with the silk, leaving the rod
positively charged by default. Recall that an electron is a tiny negatively
charged particle. Its motion in a copper wire in your house constitutes an
electric current, and electrons form the outer portions of an atom. The
positive charges in an atom arise from the protons in the atomic nucleus. In

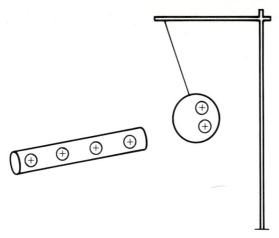

Figure 5.1 Producing an electric force. Both the glass rod and the foil-covered ball are positively charged (because each has fewer electrons than protons). The positive charges are represented by circles with a plus sign inside.

the glass rod's initial neutral state, the number of electrons equalled the number of protons. When some electrons are removed, the rod has more protons than electrons and exhibits a net positive charge.

In this chapter and the next two, all electric charges arise either from electrons (which have a negative charge) or from protons (which have a positive charge). Those charges are equal in magnitude but opposite in sign.

Figure 5.1 shows a styrofoam ball, covered with aluminum foil and suspended by a silk thread. If we momentarily touch the metal foil with the positively charged glass rod, some of the foil's electrons (which are attracted by the rod's positive charge) leave the foil, attach themselves to the rod, and reduce the rod's net positive charge. Now the ball is positively charged (by default), and the rod remains positively charged (at least somewhat so). In short, we have two positively charged objects. And we find that now we can push the ball around *without touching it*. Holding the positively charged rod first on one side of the ball and then on the other, we can make the ball swing to and fro, as though it were a child on a swing. Or we can make the ball cruise in a horizontal circle, tethered by the silk thread. With the charged rod, we are able to exert a force on the charged ball. (Recall that, in physics, a *force* is a push or a pull.)

The preceding paragraph gives us the entire scene at once. Now we begin to isolate pieces and describe them separately.

First we study the charged rod in isolation. Figure 5.2 shows the rod. Let us imagine taking a separate positive charge, holding it still at a specific point in space near the rod, and measuring the electric force exerted on it.

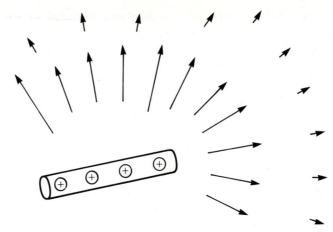

Figure 5.2 A positively charged rod and the electric field in its vicinity.

Then, at the corresponding location in the figure, we draw an arrow pointing in the direction of the force we measured. The length of the arrow is drawn proportional to the magnitude of the force. After we have done this for many locations – some close to the rod and others farther away – the picture in figure 5.2 emerges.

Next, physics says that this set of arrows is a picture of something really existing there in space. That something is called the *electric field*. The word "field" means that an arrow is associated with each point in space. Here is an analogy. Imagine a wind-blown field of wheat on a gusty day. In some places, the wheat stalks are aligned in a northward direction; in other places, the stalks have been blown down toward the southeast, and so on. At every location in the farmer's field, there is a wheat stalk, and it points in some direction. Thus, in an analogy, each arrow in figure 5.2 could represent a real physical wheat stalk. But the objective of the figure is not to describe wheat fields. Rather, each arrow *represents* the electric field at the location of the arrow's tail. The next few paragraphs take us from this beginning to a precise definition.

We generated figure 5.2 by measuring the force on an extra charge, and that very process provides a definition of the electric field. In words alone, the definition is this:

$$\text{Value of electric field at point P} = \frac{\text{force on charge we place at P}}{\text{amount of charge we use}},$$

provided the charge we use is at rest at P. Why the division? If we were to use a charge three times as large, we would find three times as much force.

To get a quantity that is independent of how much charge we use, we need to divide the force by the amount of charge.

Some symbols will be useful. A force has a direction as well as a magnitude. When we need to display the directional aspect (as well as the magnitude), we denote a force by the boldface letter **F**. The electric field at a point also has a direction as well as a magnitude, and so its symbol is the boldface letter **E**. For the amount of charge in the present context, we use the symbol q. The symbol q provides also a name for a specific charge, when we need to refer to the charge. (The next paragraph shows both uses of the letter q: amount of charge and name.)

The verbal definition of the electric field given above becomes this:

$$\text{Value of electric field at point P} = \mathbf{E} = \frac{\mathbf{F}_{on\,q}}{q} \qquad (1)$$

when charge q is at rest at P. The subscript "on q" tells us that the force $\mathbf{F}_{on\,q}$ acts on charge q (and here the symbol q serves as a name for the electric charge). In the denominator, the symbol q denotes the amount of charge on "charge q." Figure 5.3 shows the definition in action.

The idea of the electric field can be an elusive concept, but at least our definition has a notable merit: it is an *operational definition*. The adjective "operational" means that the definition is couched in terms of acts – operations – that can be performed in the laboratory or elsewhere. While one may have a hard time grasping what the electric field *is*, at least one can determine its value at a point by explicit experiment.

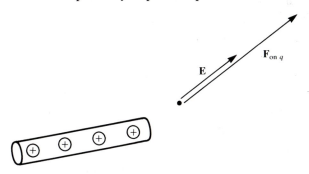

Figure 5.3 Defining the electric field. Here the charge q has a value of 2 charge units. Therefore $\mathbf{E}=(\mathbf{F}_{on\,q})/2$, and the electric field arrow is half as long as the force arrow.

Conceptual framework

Now we return to figure 5.1: the charged rod and the charged ball. The ball experiences a force. That is the obvious fact. Physics, however, conceives of that force as arising in a two-step process.

(1) The first charge (on the rod) alters the real physical properties of space everywhere: it sets up an "electric field" in space. To be sure, the first charge's influence on the properties of space diminishes with distance; yet the first charge "adds" something to otherwise empty space.

(2) The second charge (on the foil-covered ball) is acted on by the real electric field – the alteration of the properties of space – in its immediate vicinity. Thus the second charge is acted on only by its immediate surroundings.

This two-step conception elevates the electric field to the status of something real, something existing in space.

Why the field concept?

Why should physics want to introduce the idea of an electric field as an element of physical reality?

In response, let us consider radio transmission between the earth and a planetary probe, as sketched in figure 5.4. It takes about 30 minutes for a radio message to go from the earth to the vicinity of Jupiter. Radio waves are produced by electric charges moving in an antenna. We can wiggle the charges, get the message underway, and then hold the charges at rest in such a way that the antenna is electrically neutral. The radio message keeps on going. There must be something real out there in inter-planetary space, something that bears the message. Part of the something is a non-zero electric field. The field concept provides a real physical carrier for the radio message, a "something" that travels through what is otherwise empty space.

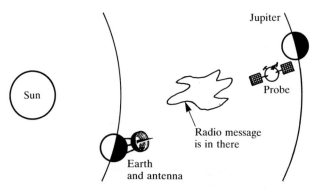

Figure 5.4 A radio message en route from the earth to the vicinity of Jupiter. The scene is set 15 minutes after a short message has been sent. The antenna on earth and the planetary probe are visible objects – but there must be something else, too.

5.2 Franklin's clapper

A more mundane example of an electric field is in order, too. Imagine two thin brass disks, their diameters being about the length of your hand. As figure 5.5 indicates, the disks are arranged to face each other and are separated by a few centimeters. The left-hand disk is charged positively; the right-hand disk, negatively. With such charges around, there must be an electric field. In which direction does it point?

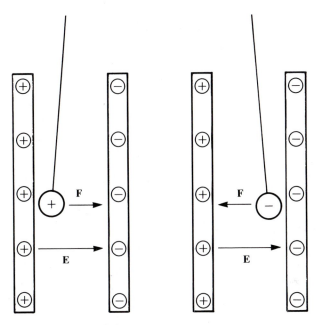

Figure 5.5 Franklin's clapper. The brass disks are seen edge-on; they are held up by posts that do not conduct electricity (and that are omitted in the sketch). Positive and negative charges are represented by circles with a plus and a minus sign inside, respectively. The iron nut, hung from a silk thread, is pushed to and fro between the charged disks. (A Wimshurst machine will charge up the disks nicely.)

Suppose we take an extra positive charge q and place it somewhere between the two disks. The positive charges on the left-hand disk will repel q toward the right. The negative charges on the right-hand disk will attract q to the right. The two forces add to give a bigger force to the right. The definition of the electric field $\mathbf{E}$,

$$\mathbf{E} = \frac{\mathbf{F}_{on\,q}}{q},$$ (1)

111

implies that **E** points to the right. This orientation will hold everywhere in the space between the two disks.

Now we dangle an iron nut, suspended by a silk thread, into the central region. Initially, the nut is electrically neutral, but we can charge it by letting it touch the left-hand disk. Some electrons are attracted from the nut to the disk, and the nut becomes positive by default.

We can use the electric field concept to figure out the force that now acts on the iron nut. If we multiply equation (1) above by q on both sides and then cancel on the right-hand side, we find

$$q\mathbf{E} = \mathbf{F}_{on\ q}. \tag{2}$$

The significance of the new equation is this: once we know the electric field **E**, we can compute the force on *any* charge q by multiplying **E** by the amount of charge q. At the moment, q is positive; **E** points to the right; and so the force on the iron nut points to the right. The nut is pushed over to the right-hand disk.

When the nut arrives, however, electrons swarm from the disk, neutralize the once-positive nut, and then overdo things, making the nut negatively charged. In equation (2), the amount of charge q is now a negative quantity; the electric field **E** still points to the right; but the interpretation of the product $q\mathbf{E}$ (with q negative) is that the force points in the opposite direction: toward the left. The nut is shoved back to where it started.

When the nut arrives at the left-hand disk, electrons will leave the nut, and the nut will become positive once more. The stage is set for another swing to the right and then back. And on and on, so long as the disks remain highly charged.

Although one still cannot literally see the electric field, one can hear its effect: each time the iron nut swings over to a disk, it hits with a tiny "clink." Benjamin Franklin invented this simple noise maker, and it carries the name "Franklin's clapper."

5.3 The magnetic field

Most of us are familiar with magnetism, at least in an informal way. Cartoons and shopping lists decorate the refrigerator door, held there by little magnets. A compass needle tells the hiker the direction of north. More precisely, the needle tells the direction of the magnetic north pole, located in the Canadian high arctic, but south of the rotational north pole. Figure

Figure 5.6 How a compass needle points at various locations on the earth's surface. Two poles are shown: the true north pole, about which the earth rotates, and the magnetic north pole, whose location is marked by the triangle. The short arrows indicate how a compass needle points; they become a map of the earth's magnetic field.

5.6 sketches a portion of the earth's surface and shows how a compass needle points at various locations.

At each location on the earth's surface, the earth's magnetism orients a compass needle in a specific direction. The needle swings but settles down to a specific direction, reproducibly. From the perspective of someone above the earth and looking at it, the needle direction varies greatly from Alaska to Florida but does so smoothly.

In short, the earth's magnetic effect on a compass needle varies from place to place, and so physics introduces the idea of a *magnetic field*.

The direction of the magnetic field is specified operationally to be the direction in which a compass needle lines up. Certainly this definition serves to determine the direction of the earth's magnetic field, and it suffices (for our purposes) in other contexts, too. Moreover, a compass needle can be our detector of magnetic fields.

The symbol for the magnetic field is the boldface letter **B**. This is the magnetic analog of **E** for the electric field. (Unfortunately, the symbol **B** provides no mnemonic aid. The choice of symbol is an alphabetical accident of history. When James Clerk Maxwell introduced a comprehensive set of symbols for electricity and magnetism, the letter **M** had already been claimed by another physical quantity. Because the magnetic field came second on the list of quantities yet to be given symbols, it received the second letter of the alphabet: **B**.)

The value of the electric field at a point has both a direction and a magnitude. So, too, does the value of the magnetic field, but we have defined the field's direction only. The reason for the omission is this: an operational definition of magnitude is technically complex; we do not need the magnitude; and so we are best off ignoring it. But some further experience with magnetism is certainly in order.

The Oersted experiment

A car battery can produce a steady flow of electrons through a copper wire connected to its terminals (at least for intervals of a few seconds). The flow of charge constitutes an electric current. Figure 5.7 sketches that context

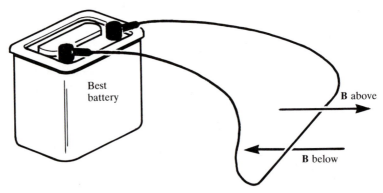

Figure 5.7 The Oersted experiment: a steady electric current produces a magnetic field. The compass needle points one way below the current-carrying wire and in the opposite direction above the wire.

A note about the electric current. When the copper wire is connected to the battery, electrons already in the wire – electrons that come from the copper atoms – begin to flow through the wire. As some of those electrons enter the battery through one terminal, other electrons emerge from the battery at the other terminal and enter the wire. The wire started off electrically neutral (but containing both electrons and positive copper nuclei); it remains electrically neutral. Thus the deflection of the compass needle is *not* a consequence of electric forces. Rather, because electrons move through the wire, there is a flow of electric charge and hence an electric current. That current is responsible for the deflection of the compass needle, and we infer the existence of a magnetic field as a real, physical intermediary.

and also shows how a compass needle responds. When placed above a straight section of the current-carrying wire, the needle points one way; when placed directly below, the needle swings around and points the other way. If only the earth's magnetic field were present, the needle would point the same way in both locations. Here we have clear evidence that electric charges in steady motion – a steady electric current – produce a magnetic field.

By 1820, the Danish physicist Hans Christian Oersted had long been looking for a connection between electricity and magnetism. While preparing one of his lectures, he thought of an experimental arrangement of electric current and compass needle that might reveal a connection. He assembled the apparatus but did not have time to try it out before class. As he lectured, his conviction grew – yes, the experiment would show a connection – and so Oersted tried the experiment right there in class. The compass needle responded to the electric current, and so the experiment sketched in this section carries the name "the Oersted experiment." It was the first experiment to establish a connection between electricity and magnetism. (Professor Oersted had only a weak source of electric current, and so the needle's response was feeble, indeed, barely perceptible. In his words, "the experiment made no strong impression on the audience." Moral: regardless of whether they are innovative or shopworn, lecture demonstrations need to be large, visible, and convincing.)

Magnetic fields produced by electric currents are all around us, but we are usually unaware of them. For example, the earth's magnetic field is believed to be produced by electric currents deep in the earth's molten interior. Scrap metal and old cars are picked up in a junk yard with an electromagnet: a magnet produced by an electric current. To drop the junk, the crane operator just turns off the current.

Magnetic forces

An electric current, we have learned, produces a magnetic field. Does a reciprocal effect occur, that is, does a magnetic field affect an electric current?

The answer is yes. Perhaps you have sat in front of a computer terminal and watched the letters and numerals appear on the screen. You may know that an electron beam (inside the TV-like tube) produces the light by impact on the inside of the tube face. But what steers the beam around from one letter to another? Or through the pen strokes of the letter A? Two electromagnets, buried deep inside the tube, perform those tasks. Their magnetic fields push on the electrons and deflect the beam, steering it from place to place on the tube face.

5.4 The entities

A pause to summarize is in order. What are the entities that we have met in this chapter?

- Electric charge. Electrons and protons are tiny particles of matter. The electron has a negative charge, and the proton, a positive charge. (The hydrogen atom – the simplest of all atoms – consists of merely an electron and a proton. The two charged particles attract each other and so stay near each other, forming an atom.)
- Electric current. An electric current is nothing but electric charge in motion. The flow of electrons in a copper or aluminum wire is what carries "electricity" around your house.
- The electric field **E**.
- The magnetic field **B**.

But what, you may ask, *are* the electric and magnetic fields? They are defined operationally in terms of forces on charged particles or compass needles. They can exist in what is otherwise vacuum (as in the instance of a radio message sent to a planetary probe near Jupiter). The fields are fundamental; I cannot express them in terms of anything more primitive or more intuitive. We can visualize the pattern they make in space, but we cannot "see" them literally.

In our conceptual framework, the fields are the agents immediately responsible for electric and magnetic forces. For example, the foil-covered ball in figure 5.1 responds to the electric field in its vicinity and experiences an electric force. To be sure, the charged rod produces the electric field, but physics insists on the electric field as an intermediary between the charged rod and the charged ball.

5.5 Faraday's law

We just noted that electric and magnetic fields exert forces on charged particles. Physics takes the field idea very seriously, and so we are led to ask, can one kind of field have an effect on the other field? Specifically, can one kind of field produce the other? (In the contexts that we have studied

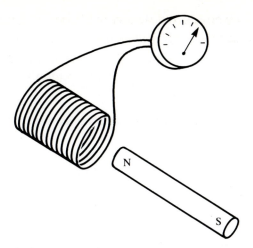

Figure 5.8 Moving the bar magnet relative to the coil of wire causes the needle of the current meter to deflect. Inference: when a magnetic field changes with time, an electric field is produced.

thus far, the fields have been produced by electric charges, at rest or in motion.)

Figure 5.8 sketches apparatus with which we ask, experimentally, can a magnetic field produce an electric field? A coil of wire is connected to a sensitive current meter. At the start, no current flows, and the meter reads zero. Only if an electric field arises within the wire will current flow; then the electrons will be pushed around the circuit consisting of wire and meter. Thus a non-zero meter reading indicates the presence of an electric field.

A bar magnet provides a convenient source of magnetic field. The magnet produces a strong field in its vicinity and not much field far away.

When we thrust the bar magnet into the coil of wires, the current meter responds, indicating a flow of electrons. We infer that an electric field has arisen. But the current quickly drops back to zero. The magnet is sitting inside the coil; there is plenty of magnetic field inside and around the coil; but there is no longer a current and hence no longer an electric field.

We remove the magnet from the coil – and again the meter responds, but only momentarily.

A current arises when the magnet is being inserted or withdrawn, but only then. During those periods, an electric field is produced. And during those periods, the magnetic field at the location of the coil is changing, either from zero to some large value or from the large value back to zero. Our inference is this: a *changing* magnetic field generates an electric field.

A host of other experiments – some of which you may see – confirm this conclusion. The name for the conclusion, *Faraday's law*, honors the English physicist Michael Faraday, who discovered the effect in the 1830s. The son of a blacksmith and originally apprenticed as a bookbinder, Faraday rose by his industry and skill to become director of London's Royal Institution.

5.6 Electromagnetic waves

In the last section, we learned that a changing magnetic field generates an electric field. Is there reciprocity? Does a changing electric field generate a magnetic field?

The answer is yes, but here a simple experiment is not possible. Rather, we need to do some reasoning and approach the question indirectly.

Here is the logic, expressed tentatively. If

> a changing magnetic field generates an electric field and
>
> a changing electric field generates a magnetic field,

then we may be able to have both fields changing and generating each other. To be sure, we would need some electric charges to produce the fields in the first place, but once we got the show going, the fields would sustain themselves.

To peek ahead, if this conjecture is borne out, we can understand radio transmission to the Jupiter probe.

Figure 5.9 shows a device for producing self-sustaining electric and

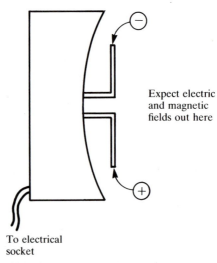

Figure 5.9 An antenna for producing self-sustaining electric and magnetic fields. Electrons are sent up and down the vertical wires periodically.

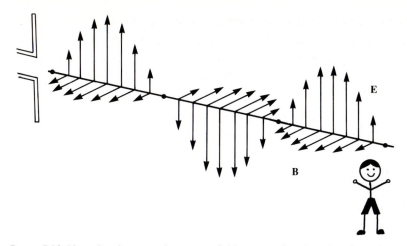

Figure 5.10 How the electric and magnetic fields appear. For the sake of our spatial orientation, the antenna wires are shown on the far left, but the fields close to the wires have been omitted from the sketch. The field values are given solely for locations along the line extending rightward from the antenna's center. Similar patterns of **E** and **B** exist along other lines emanating from the antenna.

magnetic fields. Its essence consists of two vertical pieces of wire and internal circuitry to send electrons up and down those wires. When the electrons have been sent up, the upper wire is negatively charged and the lower wire, positively charged, as the sketch shows. A moment later, the electrons will be sent down; the lower wire will be negatively charged, and the upper wire, positively charged. Then the electrons will be sent back to the upper wire and so on, periodically.

The motion of the electrons constitutes a current. We can expect that current to produce a magnetic field. When one wire is positively charged and the other negatively, each wire will make an electric field, and those electric fields will not cancel each other. So we can expect, typically, a non-zero electric field. While this is not the whole story for a periodically-varying distribution of electric charge, it certainly suggests that the device will generate both electric and magnetic fields. Optimistically, we call it the transmitting antenna.

Figure 5.10 gives a perspective drawing of the fields that we can expect. Because the electrons are driven up and down the wires periodically, the fields show a periodic pattern. The orientation of the magnetic field is similar to what figure 5.7, the Oersted experiment, showed for the field produced by a steady current: an orientation perpendicular to the wire. The electric field is lined up parallel to the way the charges are separated on the antenna.

If this pattern of fields moves through space (along the line indicated), a stationary observer – his name is Bob – notes temporal variation. As the pattern passes Bob, first the electric field points up; then it decreases and soon points down; next it increases and again points up. Similar behavior, but tipped by 90 degrees, holds for the magnetic field. Such changes in time (at a fixed location) provide the "changing electric field" and the "changing magnetic field" that generate the magnetic and electric fields, respectively, and keep the show going even after the antenna has been unplugged.

But how much of this is speculation and how much real? The circuitry in the transmitting antenna can easily send the electrons up and down the wires with a frequency of oscillation equal to 3×10^9 round-trips per second. At a distance of about 4 meters, one can detect the electric field by letting it push electrons through another wire, this one connected to a current meter. The presence of a current implies the existence of an electric field to urge the electrons along. The combination of wire and meter is our detector of the electric field.

What does one find?

First, when we place the detector out where figures 5.9 and 5.10 suggest that an electric field exists, the electric field is indeed there.

Second, when a large sheet of aluminum, 1 millimeter thick, is placed between the antenna and the detector, the metal reads zero current and implies no electric field at the detector location. But figure 5.11 reveals what is going on. If the aluminum sheet is tilted relative to the line from the antenna to the old detector location, then at the place labeled "new location" the detector does find an electric field. (No field was present there originally.) The pattern of electric and magnetic fields has been reflected. Indeed, one can hold the detector fixed at the new location and rotate the aluminum sheet back and forth a bit (about a vertical axis running through the sheet), alternately causing the reflected pattern to hit the stationary detector and to miss it.

Third, if one uses an aluminum sheet with two "slits," each a rectangle 5 centimeters by 15 centimeters, one has the analog of the double-slit experiment with light. With the antenna as source, the experiment is not easy (because stray reflections from walls confuse the pattern behind the aluminum sheet), but a double-slit interference pattern can be found.

Other experiments are possible. For example, a giant prism made of paraffin wax produces refraction of the traveling pattern.

We may take these experiments as good evidence that electric and magnetic fields are able to sustain each other. Once started, the fields move through space on their own, propagating in a wave-like fashion.

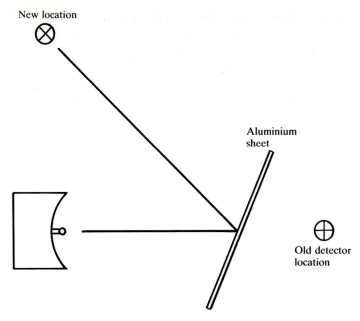

Figure 5.11 A top view that illustrates reflection of the traveling pattern of electric and magnetic fields. In this configuration of antenna and aluminum sheet, electric field is no longer found at the "old detector location," but electric field is observed at the "new location."

The traveling pattern in figure 5.10 is one example of an *electromagnetic wave*. The direction in which the pattern moves is called the *propagation direction*. Both the electric field and the magnetic field are perpendicular to the propagation direction. In chapter 3, we studied waves on a metal skeleton; the motion of the ribs that form the wave pattern is perpendicular to the motion of the wave crests and troughs. We called such waves transverse waves. Electromagnetic waves are formed of electric and magnetic fields that are perpendicular to the propagation direction; so it is appropriate to say that electromagnetic waves are transverse waves, too.

James Clerk Maxwell

The hero of this chapter is James Clerk Maxwell. A Scottish physicist, Maxwell studied Faraday's *Experimental Researches* and made himself master of all that was known about electricity and magnetism in the 1850s (the decade before the American Civil War). He sought a single, unified description of these two broad classes of phenomena. Success came in 1864, and Maxwell's paper, "A Dynamical Theory of the Electromagnetic Field," appeared in the *Philosophical Transactions* of the Royal Society in 1865.

Not only was there unification of the phenomena, but Maxwell's theory predicted self-sustaining and propagating electric and magnetic fields – what we have called electromagnetic waves.

Maxwell was able to predict the speed at which such waves should travel: 3×10^8 meters/second. The experimental data used in this prediction came from a set of strictly electric and magnetic measurements on charges at rest or in motion in wires. No experiments on light entered into the prediction. As Maxwell remarked, "The only use made of light in the experiment was to see the instruments" and to read them. Yet we recognize the predicted speed as coinciding with the speed of light, known since Newton's day. Maxwell was well prepared to recognize the coincidence, too. A link between light on the one hand and electricity and magnetism on the other had been sought for decades, especially by Faraday, and with a modicum of success. And so no one is surprised to find Maxwell writing as follows:

> This velocity is so nearly that of light [when the comparison is made to more decimal places] that it seems we have strong reason to conclude that light itself (including radiant heat and other radiations, if any) is an electromagnetic disturbance in the form of waves propagated through the electromagnetic field according to electromagnetic laws.

No surprise, but the statement is staggering in its implications: *light is an electromagnetic wave*.

5.7 The electromagnetic spectrum

To say that light is an electromagnetic wave means this: the red beam from a neon laser consists of a traveling pattern of electric and magnetic fields. In chapters 3 and 4, experiments told us that light is a wave phenomenon, but those chapters left unanswered the question, what is the "something" out of which light waves are formed? Now we have an answer: light waves are formed out of electric and magnetic fields. The fields are the "something." When light streaks invisibly from the laser to a red spot on the wall, the intervening space contains electric and magnetic fields, a wave-like pattern of them – indeed, a pattern traveling at the speed of light.

The wavelength of that pattern we already know: $\lambda_{\text{neon red}} = 6.3 \times 10^{-7}$ meters. This is illustrated in figure 5.12. We can imagine an observer stationed somewhere along the beam. The pattern travels by at the speed c; so high a speed and so short a wavelength send crest after crest past the

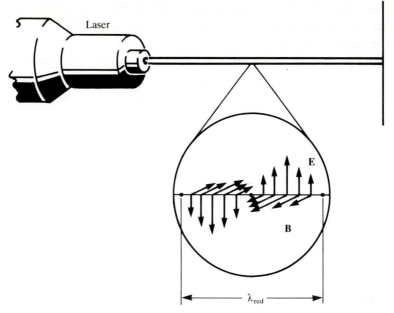

Figure 5.12 When a neon laser sends a continous light beam to the distant wall, electric and magnetic fields fill an approximately cylindrical volume of space, the space occupied by the beam. (The cylinder is about 1 millimeter in diameter.) An enlarged rendition of a tiny section reveals the fields and illustrates the wavelength.

observer at an astonishingly high frequency. Our general relationship among frequency, wavelength, and speed is

$$f \lambda = v.$$

Specializing this equation to light, we have

$$f \lambda = c. \tag{1}$$

To diplay the frequency f in isolation, we divide both sides by λ:

$$f = \frac{c}{\lambda}.$$

And so, for the neon beam, we find

$$f_{\text{neon red}} = \frac{c}{\lambda_{\text{neon red}}}$$

$$= \frac{3 \times 10^8 \text{ meters / second}}{6.3 \times 10^{-7} \text{ meters}}$$

$$\approx \tfrac{1}{2} \times 10^{15} = 5 \times 10^{14} \text{ oscillations / second.}$$

At a fixed location, the electric field changes from up to down and back to up about 5×10^{14} times each second.

The transmission antenna of section 5.6 sent electrons up and down the vertical wire at a frequency of $f=3\times10^9$ oscillations/second. The traveling pattern is repeated at this frequency. What is the pattern's wavelength? Now we need to isolate λ in equation (1); division on both sides by f does the job:

$$\lambda = \frac{c}{f}.$$

And we find

$$\lambda = \frac{3\times10^8}{3\times10^9} = 0.1 \text{ meters}.$$

The antenna sends out waves with a wavelength of 10 centimeters, roughly the width of your hand.

We have now two instances of electromagnetic waves: the 10-centimeter waves and the neon laser beam. They are but two points along an entire spectrum of possibilities. Nature provides us with electromagnetic waves having other frequencies and wavelengths, and technology enables us to produce some at will. Figure 5.13 shows major ranges in the *electromagnetic spectrum*. Visible light occupies a narrow interval, bounded by infrared

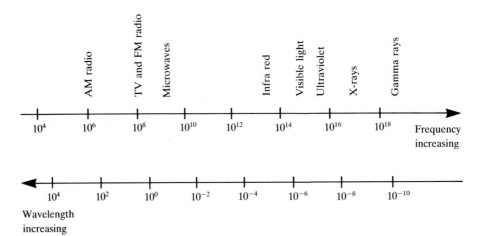

Figure 5.13 The electromagnetic spectrum. The numerical values of frequency are given in oscillations per second; those of wavelength, in meters. Nature provides a continuous spectrum, but physicists have divided the spectrum up into intervals, some narrow but others broad. All the types of radiation noted here share certain essential features: they have a wave-like character, are formed from electric and magnetic fields, and travel (in vacuum) at the speed of light.

radiation (starting at frequencies below those of red light) and by ultraviolet light (starting at frequencies higher than those of violet light). The 10-centimeter waves of the last section are an instance of microwaves. The electromagnetic waves that heat a croissant in a microwave oven have, typically, a frequency of 2.45×10^9 oscillations/second and hence a wavelength of 12.2 centimeters. If you roast a turkey at 350 degress Fahrenheit in a conventional oven, the radiation filling the oven is predominantly infrared. (The skin of your hand is a good detector of that radiation, but the retina of your eye is not.) If you tune in channel 3, your television set searches for electromagnetic waves at a frequency of 61.25×10^6 oscillations/second. The associated wavelength is $\lambda = c/f = 5$ meters. Classical music from an FM radio station at 88.6 FM is carried by electromagnetic waves whose underlying frequency is 88.6×10^6 oscillations/second. Thus TV and FM radio broadcast at similar frequencies. If, however, you are an aficionado of old-fashioned AM radio, then 1080 on your dial brings in a message sent from the broadcasting station on waves at a frequency of 1080×10^3 oscillations/second, a much lower frequency. The associated wavelength is huge: about 300 meters. When police use radar to monitor the speed of highway traffic, an electromagnetic wave travels from the cruiser, reflects off the speeding car, and returns to an antenna on the cruiser. The wavelength of such a radar beam is a few centimeters (anywhere from 1 to 5 centimeters, really) and places the radiation in the microwave interval.

The frequencies above the visible are less common in our lives, but certainly not absent. Light from the sun contains substantial ultraviolet light (in addition to visible and infrared). A bad sunburn results from too much ultraviolet light; you are especially vulnerable at the beach or on a ski slope, where the surface reflects the ultraviolet back for a second chance at being absorbed by your skin. When the dentist takes an X-ray of your molars, electromagnetic waves with a frequency of perhaps 1.5×10^{19} oscillations/second pass through your jaw and teeth, producing a photograph on film held in your mouth. Gamma rays are rare in our lives. Their origin is typically the nuclei of atoms undergoing radioactive decay. Just as the transmitting antenna of section 5.6 used the motion of electrons in wires to generate microwaves, so the motion of protons in the atomic nucleus can produce electromagnetic radiation, then called gamma radiation. (That electromagnetic radiation was the third in a trio of once-mysterious kinds of radiation, labeled *alpha*, *beta*, and *gamma*, merely the first three letters of the Greek alphabet. The three kinds of radiation were lettered in order of increasing ability to penetrate matter; the gamma rays were the most difficult to stop. The alpha radiation turned out to be just a compact cluster of two protons and two neutrons; the beta type turned out to be ordinary electrons; only the third kind – gamma radiation – is electromagnetic in character.)

5.8 "Sidedness" explained and explored

Back in chapter 1, experiments with clip-on sunglasses or their equivalent led us to ascribe some kind of "sidedness" to light. Now we can make theoretical sense of this. In figure 5.10, the electromagnetic wave has the electric field arrayed in one plane and the magnetic field in another plane, perpendicular to the first. In colloquial language, the wave has electric field pointing out on two sides (up and down in the figure) and magnetic field pointing out on two other sides (in and out of the page). The fields give a "sidedness" to an electromagnetic wave and hence to light. The "sides" of the light wave that have **E** pointing out differ from the "sides" that have **B** pointing out.

In what follows, we concentrate our attention on the electric field. The magnetic field is always present, too, but working with one field is sufficient.

First, what happens when light impinges on a sheet of Polaroid? Recall that the sheet consists of stretched and partially-aligned strands of polyvinyl alcohol, together with iodine atoms that string themselves along each molecular chain. Figure 5.14 depicts the fate of light whose electric field is either wholly parallel to the stretch direction or wholly perpendicular: the light is either wholly absorbed or wholly transmitted. (No sheet does this perfectly, but such is the ideal.) An analogy can help us to understand why this happens.

The iodine-coated strands act somewhat like electrically conducting

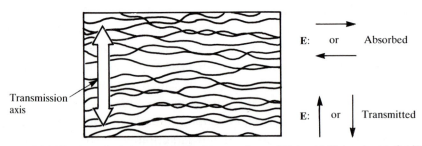

Figure 5.14 The fate of a light wave impinging on a sheet of Polaroid. If the electric field is parallel to the stretch direction, the wave is absorbed; if perpendicular, then transmitted. The broad, double-headed arrow denotes the *transmission axis*.

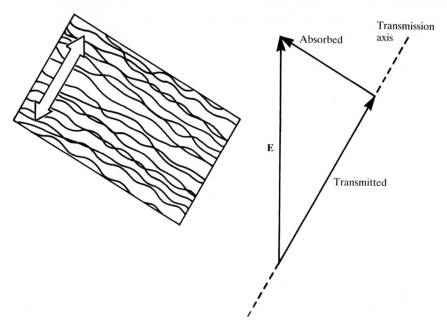

Figure 5.15 What happens when the incident electric field is oblique to the transmission axis.

wires. To be sure, not like very good conductors, but like conductors nonetheless. If the electric field is parallel to the "wire," some electrons are pushed along; there is microscopic current and heating (as when current flows through the wires of an electric toaster). The energy for this heating comes from the electromagnetic wave, and so the light is absorbed. If, however, the electric field is perpendicular to the "wires," the field cannot push the electrons far; there is no current, no heating, and no absorption.

Remarkably, figure 5.14 contains all the information about Polaroid sheets that we need. If the electric field is oblique to the transmission axis, as shown in figure 5.15, we decompose the field into components that are wholly parallel and wholly perpendicular to the axis, as indicated. The parallel component is transmitted; the perpendicular, absorbed. Some light will emerge, but it will be dimmer in intensity. And, of course, the emergent light will have its electric field purely along the direction of the Polaroid's transmission axis.

The last sentence has several ramifications. Light whose electric field always oscillates in the same plane is called *plane polarized* light. The electromagnetic wave in figure 5.10 is plane polarized, for the electric field always points either "up" or "down" in the plane of the page. The light from a hot filament comes from many atoms and their individual electrons.

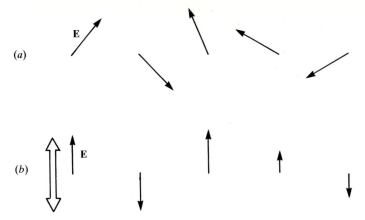

Figure 5.16 Making plane polarized light. (*a*) At a fixed point, the electric field of light from a hot filament points first one way, then another, yet a third, and so on. The sequence shows the field at different instants of time (but at the same location). (*b*) After the light has passed through a Polaroid sheet, only the components parallel to the transmission axis remain.

The lengths of the arrows vary, and so you might wonder whether our eyes would notice some temporal variation in the intensity of the transmitted light. The answer is "no." The changes follow one another so rapidly that our eyes respond only to an average over the lengths of many arrows.

The ensuing electric field is a combination of many unrelated contributions. Just what those contributions add up to varies significantly from moment to moment. The field does not have a fixed plane of oscillation but rather oscillates for a while this way, then that way, and so on. Figure 5.16 illustrates the situation. If we put a sheet of Polaroid in front of the filament, then only the component of **E** parallel to the transmission axis will be allowed to pass. The perpendicular component is absorbed. The transmitted light is dimmer but plane polarized – and a good light source for an experiment.

Another experiment

Indeed, let us put together a last experiment for this chapter. In front of a frosted light bulb with its hot filament, we place a Polaroid sheet with the transmission axis vertical. In front of that sheet we place another sheet, but with its transmission axis horizontal. Between the two sheets is plenty of light, its electric field oscillating in the vertical plane, but none of that light gets through the last Polaroid sheet. Because the electric field of the "in between" light is perpendicular to the last sheet's transmission axis, the last sheet absorbs it all.

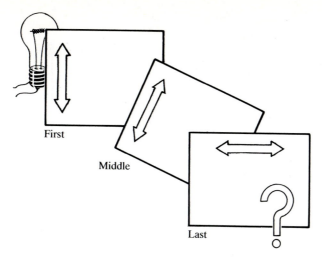

Figure 5.17 Stacking three Polaroid sheets. The last sheet has its transmission axis perpendicular to the first sheet's transmission axis. The middle sheet has its transmission axis oriented obliquely relative to the axes of the other sheets. Can any light run this gauntlet?

Nothing about that is surprising, but now suppose we insert a middle Polaroid sheet between the first and last and do so at some oblique angle. The transmission axis might be tilted 30 degrees relative to the first sheet's transmission axis. Figure 5.17 provides a sketch. We have inserted another absorber of light. But it is a selective absorber, absorbing only the component of electric field perpendicular to its transmission axis. Before we jump to the obvious conclusion – that no light could possibly get through this sequence of absorbers – let us work our way along, thinking about components.

After light from the frosted bulb passes through the first sheet, the light's electric field is purely vertical. When light reaches the middle sheet, as sketched in part (a) of figure 5.18, the electric field component along that sheet's transmission axis is transmitted. Thus the light traveling from the middle sheet to the last sheet is plane polarized in an oblique direction. When that light reaches the last sheet, its field component parallel to the transmission axis will be transmitted. As the diagrams show, the portion transmitted is not zero. In short, we predict that light will now emerge – dimmer, certainly, than light coming directly from the bulb, but light nonetheless.

The experiment is easy to set up – and it confirms the prediction. The sidedness of light makes it possible for more absorbers to produce less absorption!

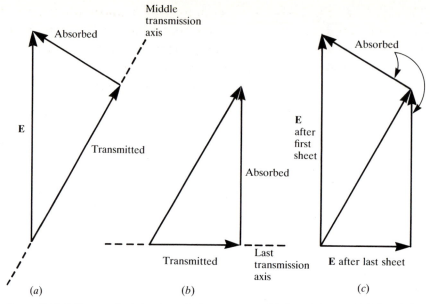

Figure 5.18 Following the light through the sequence of Polaroid sheets. (a) When light impinges on the middle sheet. (b) When light reaches the last sheet. (c) A composite of what happens at the middle and last sheets.

5.9 Odds and ends

Heinrich Hertz

It is time to set some history straight. In electromagnetism, we have two fields: electric and magnetic. Our study led us to a conjecture: when changing in time, each field will induce the other field. A description of lecture demonstrations provided experimental support for that idea. Then James Clerk Maxwell entered our development. In 1864, Maxwell predicted that electromagnetic waves exist and travel at the same speed as light. He went on to suggest that light itself is an electromagnetic wave. We adopted this view and examined the great spectrum of electromagnetic waves, a spectrum in which (visible) light occupies an interval – but only a small one. This is a good sequence of events when one is trying to grasp a subject as difficult as ours, but the historical development was different. Let us examine it.

When Maxwell predicted the existence of electromagnetic waves in 1864, no one had ever moved electric charges around in a laboratory and deliberately produced electromagnetic waves. His "prediction" of such waves was a *bona fide* prediction, a prophesy, a foretelling. And the identification of light as an electromagnetic wave was likewise a prediction, to be tested experimentally.

The world of physics was keenly interested in Maxwell's theory, but experimental confirmation pre-supposes apparatus capable of performing the tests. Or it pre-supposes an individual so talented as to be able to devise new apparatus. In 1864, the technical repertoire of physics could not meet the challenge of testing Maxwell's predictions. More than twenty years had to pass. Finally, in the years 1886–8, Heinrich Hertz succeeded.

In those years, Hertz was professor of physics in Karlsruhe, Germany. His transmitter of electromagnetic waves was fundamentally like ours of section 5.6: electric charges were sent up and down two vertical wires. For a detector of such waves, Hertz used the spark which an electric field produces when it tries to push electric charges across a tiny gap in another wire. The spark was minute, only 1/100 of a millimeter in length, and visible only to a dark-adapted eye in a perfectly dark room. Even twenty years after Maxwell's fundamental paper appeared, experimental test was just barely achievable.

Hertz was able to discern sparks as far as 16 meters from his transmitter. That was as far as his laboratory space allowed him to go, and it was amply far enough to show him that an electric field was propagated in a beam-like fashion, like the straight-line motion of light (in a single uniform substance).

With a large sheet of the metal zinc (for aluminum was not readily available in those days), Hertz could reflect the beam from his transmitter. Moreover, the angles of reflection and incidence were equal. With a huge prism of solidified asphalt – half a ton of it – he caused the beam to refract.

To explore the "sidedness" of electromagnetic waves, Hertz built the predecessor of a Polaroid sheet: a human-sized wooden frame across which were stretched copper wires, forming a grid of parallel lines 3 centimeters apart. When the grid wires were aligned parallel to the vertical wires of the transmitter, the beam did not pass through the grid of wires. When the grid wires were perpendicular to the transmitter wires, the beam passed through freely. Now we examine the reasons for this behavior.

When the electric field is parallel to the copper grid wires, the field produces a large electric current. Because copper is such a good conductor, not much energy is literally absorbed in the wires; rather, most of the energy is reflected (just as a metal mirror reflects light waves). Thus no wave passes through the grid of copper wires. When the electric field is perpendicular to the wires, there is no place to push electric charge, and hence no reaction on the beam itself occurs: the beam passes through the grid unaffected.

Hertz's copper grid wires have the same transmitting properties that the

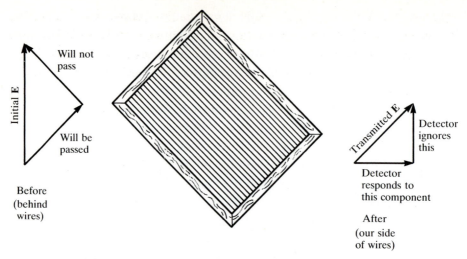

Figure 5.19 How Hertz explored the "sidedness" of his electromagnetic waves. The wooden frame holds a parallel array of copper wires; they are 1 millimeter thick and spaced 3 centimeters apart.

iodine-coated polyvinyl alcohol does in a Polaroid sheet. To be sure, this is an historical inversion: Hertz came first, and our analogy for understanding Polaroid sheets is based on his insight and experience. Moreover, in the Polaroid, it is literally absorption (not reflection) that prevents transmission when the electric field is parallel to the strands.

Hertz even performed an analog of the Polaroid experiment sketched in figure 5.17. He generated a wave with a vertically-oriented electric field, but he set his detector to respond only to the horizontal component of an electric field. Of course, the detector gave no response. Then, between transmitter and detector, Hertz interposed his grid of wires at 45 degrees. *Voilà!* The detector responded. Hertz reasoned as sketched in figure 5.19 and as described in the next paragraph.

To understand how the grid of wires affects the electromagnetic wave, decompose the initial, vertical electric field into components parallel to the wires and perpendicular to them. The parallel component will be reflected and hence will not pass through. The perpendicular component will be allowed to pass. Thus, once past the grid, the electric field is oriented at 45 degrees to the vertical. But that orientation is also 45 degrees with respect to the horizontal, and so the final electric field has a horizontal component. Now the detector responds – and establishes that the electromagnetic wave has a kind of "sidedness."

Hertz observed interference and used it to measure the wavelength of his electromagnetic waves: 66 centimeters. He had no way to measure the frequency of the waves, but he could calculate the frequency of oscillation of the electric charges in his transmitter; he based the calculation on the size and structure of the circuitry. Within some generous allowance for error here, Hertz found the product of frequency and wavelength to be about the same as the speed of light.

In the conclusion of his 1888 paper, Hertz could proudly write

> The experiments described appear to me, at any rate, eminently adapted to remove any doubt as to the identity of light, radiant heat, and electromagnetic wave motion. I believe that from now on we shall have greater confidence in making use of the advantages which this identity enables us to derive both in the study of optics and of electricity.

Newton and Maxwell

When Heinrich Hertz provided his ingenious experimental support for Maxwell's theory, he placed the capstone on the edifice of classical physics. Newton's laws of motion describe the response of a material object to a force: they describe the way the trajectory bends through space. Maxwell's laws of electromagnetism describe the mutual interaction of electric and magnetic fields and the relationship of the fields to their sources: electric charges at rest or in motion. If one throws in for good measure Newton's quantitative description of gravitation, one has the laws that constitute the fundamentals of classical physics. The laws are expressed mathematically as equations relating the relevant variables, such as velocity or electric field. A mere handful of such laws provides the theoretical basis for an astounding technology.

Table 5.1 displays some of the notable events as electromagnetic theory and its associated technology developed. The telegraph, you will note, came quite early. In telegraphy, an electric current is sent along a pair of wires, and no wave ideas are necessary for its development. Just open and close a switch, much as you can send messages in Morse code by pushing a doorbell button, thereby closing a circuit and letting electric current ring a distant bell.

The development of radar began shortly before World War II, but the war itself gave urgency to the work, in both England and the United States. The Battle of Britain was won by courage and radar. In the late 1940s, several physicists used a radar beam to measure the speed of electromagnetic waves. A half-century after Hertz began the project, they established to high accuracy that what are manifestly electromagnetic waves travel (in vacuum) at the same speed as visible light (to within the accuracy

Table 5.1. *Some notable events and dates in the history of electromagnetism and the technology that evolved from it. The classical physics of Newton and Maxwell sufficed for the engineering development of all the devices cited.*

1950	
	Commercial FM radio. Electronic computers.
1940	
	Commercial television. Radar.
1930	
1920	
1910	
1900	
	AM radio ("Wireless telegraphy").
1890	
	Hertz's confirmation of Maxwell's theory.
1880	
	Telephone. Electric lights. Phonograph.
1870	
	Maxwell's electromagnetic theory.
1860	
1850	
1840	
	Telegraph. Electric generator. Faraday's law.
1830	
	Electric motor.
1820	Oersted's experiment.

of the experiments, where the uncertainty came only in the sixth figure). Though neither Maxwell nor Hertz would have had any nagging doubts, they would have been pleased to see the results.

Electromagnetic theory: a recapitulation

Before we move on to another chapter, we would do well to summarize what this chapter has taught us about electromagnetic theory. Here are the essentials.

Physics introduces the electric field **E** as a conceptual intermediary and as a physically existing entity. One electric charge produces an electric field in the space surrounding it; that field exerts a force on a second electric charge. In this situation, the electric field is manifestly a conceptual intermediary in the production of a force, but the field also "exists" as an alteration of the properties of space. The "existence" attribute is most apparent when we consider radio transmission from earth to a planetary probe: a propagating electric field exists in otherwise-empty space and carries the message (even after the transmitter has been shut off).

The operational definition of the electric field is this:

$$\text{Value of electric field } \mathbf{E} \text{ at point P} = \frac{\text{force on charge we place at P}}{\text{amount of charge we use}}$$

$$= \frac{\mathbf{F}_{on\ q}}{q}.$$

Physics introduces also the magnetic field. Its direction is defined by the direction in which a compass needle lines up, but we did not bother to define its magnitude.

An electric charge – both when at rest and when in motion – produces an electric field. An electric charge at rest does *not* produce any magnetic effect, but an electric charge in motion – an electric current – does produce a magnetic field.

Once produced by electric charges, electric and magnetic fields can sustain themselves (in some circumstances, though not invariably). A changing magnetic field generates an electric field – that property is codified as Faraday's law – and a changing electric field generates a magnetic field.

This property of "mutual support" enables electric and magnetic fields to propagate through space. The moving pattern of fields constitutes an electromagnetic wave. Such a wave moves through vacuum at the speed of light.

Indeed, visible light itself is an electromagnetic wave. Moreover, nature provides us with an entire spectrum of electromagnetic waves. Such waves range from AM and FM radio at the low frequency, long wavelength end of the spectrum to X-rays and gamma rays at the high frequency, short wavelength end. (These are the limits of most practical concern, but waves beyond them can exist, and some do so.)

The electric field is perpendicular to the wave's propagation direction, and so is the magnetic field. Thus electromagnetic waves are transverse waves. More importantly, this perpendicularity gives electromagnetic waves a "sidedness" and explains the sidedness of light that Polaroid sheets demonstrate so dramatically.

We can see wires and batteries, but the central elements in electricity and magnetism are the fields, and they are not visible *per se*. Getting used to thinking about the fields is hard, and it was similarly so for physicists in the nineteenth century. In his little book *Relativity and Its Roots*, Banesh Hoffmann put the transformation in viewpoint vividly:

> At a time when people were asking what kept bodies moving, Galileo told them to ask, rather, what brought bodies to rest or otherwise changed their motion. Faraday initiated a comparable revolution. At a time when people were concentrating their attention on the visible electromagnetic hardware, he told them to think, rather, of the rich, invisible content of the surrounding space – the electromagnetic field.

Additional resources

The charming article "Hans Christian Oersted – Scientist, Humanist and Teacher," by J. Rud Nielsen, appears in the reprint volume *Physics History from AAPT Journals*, Melba Newell Phillips, editor (American Association of Physics Teachers, College Park, MD, 1985). The compilation contains also articles on Faraday and Hertz.

James Clerk Maxwell: A Biography, by Ivan Tolstoy (University of Chicago Press, Chicago, 1982), ably fulfills the promise of its title.

Joseph Mulligan writes engagingly about "Heinrich Hertz and the Development of Physics" in an article carrying this title in *Physics Today*, March 1989, pages 50–7.

Questions

1. (*a*) Calculate the time interval needed for one full oscillation of the electric field of yellow light. Take the wavelength of yellow light to be $\lambda_{yellow}=5.8\times10^{-7}$ meters in vacuum. (You may find it useful to calculate the frequency as an intermediate step.)

(*b*) About how many wavelengths are included in the light wave produced by a single atom when it emits a burst of light? The burst occurs over a time interval of about $0.000\,000\,001$ seconds (that is, 10^{-9} seconds).

(*c*) How long in space is the light wave, that is, the distance from the front to the end of the entire wave pattern?

Surprising numbers? Yet these numbers are characteristic of the light we get every sunny day from the sun.

2. Let's return to question 6(*b*) of chapter 1. Here it is, with one change: the phrase "particle theory" has been replaced by "wave theory."

> A triangular glass prism splits white light into a spectrum, different colors being bent through different angles. Suggest one or more ways in which a wave theory of light can explain this phenomenon and yet be consistent with a single speed of light for all colors in vacuum.

What is your suggestion now?

3. Suppose we set up a double-slit interference experiment just like the one we used to measure λ for the laser light and then put a separate Polaroid behind each slit. The light source is a filament lamp (giving white light) followed by a red filter. The apparatus is sketched in figure 5.20.

Recall that, in a beam of light from a hot filament, the electric field, though always perpendicular to the propagation direction, does not always oscillate in the same plane. The orientation swings around, as was illustrated in figure 5.16.

(*a*) What will the pattern on the screen be like if both Polaroids are oriented in the same direction (as though we used a single large sheet)?

(*b*) What will the pattern on the screen be like if one Polaroid is aligned to transmit light with **E** parallel to the slit and the other Polaroid, light with **E** perpendicular to the slit? Why? Can there be destructive interference? (You may find it helpful to draw a sketch showing the electric fields that come from the two slits.)

4. Which "bends" more, red light or violet? Consider two different physical contexts:

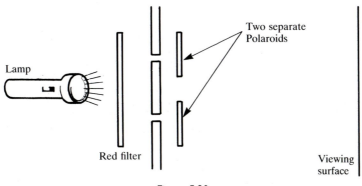

Figure 5.20

137

(1) Light going through a glass prism.

(2) Light going through a pair of closely-spaced slits.

For each context, respond to the following items.

(*a*) Which "bends" more, red light or violet?

(*b*) What is the technical term that should replace the informal notion of "bends"?

(*c*) In a few sentences, describe the essence of the theory that we use to understand the "bending."

5. The speed of red light in a certain transparent material is 2.1×10^8 meters/second. Use the wave theory in the following questions.

(*a*) For the situation in part (*a*) of figure 5.21, where a beam of red light strikes the material, calculate either the semi-chord of the refracted ray or the sine of the angle of refraction. You will need to make your own drawing of the circle, etc. On your drawing, indicate the refracted ray and its semi-chord.

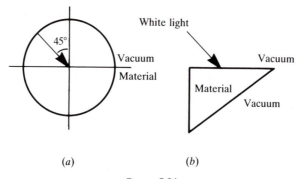

(*a*)   (*b*)

Figure 5.21

(*b*) When violet light is incident at the same angle, the semi-chord of the refracted ray is $0.45R$, where R denotes the radius of the circle. What is the speed of violet light in the material?

(*c*) If the material were cut into a prismatic shape, as shown in part (*b*) of the figure, would the material produce a normal spectrum (like that produced by glass)? The alternatives are no spectrum at all or a reversal of the order of the colors. Explain why you answered as you did.

6. Figure 5.22 shows two straight copper wires with a flashlight bulb con-

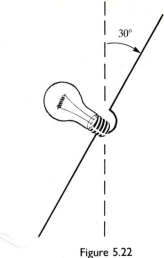

Figure 5.22

nected between them (so that an electron could move along one wire, pass through the bulb's filament, and then continue along the other wire). This combination of two wires plus bulb forms a "receiving antenna" for electromagnetic waves. The transmitting antenna is similar to that shown in figure 5.9.

If the transmitter (which has vertical antenna arms) produces an electric field **E** that oscillates vertically (first pointing straight up, then down, then up again, and so on), why does the bulb still light up (somewhat, at least) when the receiving antenna is tipped at 30 degrees, say, relative to the vertical?

3. Between Alice and the (harmless) laser is a thick sheet of plate glass, as shown in figure 5.23.

(*a*) Reproduce the sketch and then draw carefully the continuation of

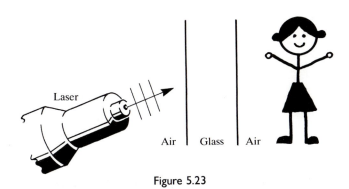

Figure 5.23

the light beam as it (or some of it) passes through the glass and then reaches Alice or fails to reach her. Aim to draw the path qualitatively faithfully.

(*b*) Include representative crest lines both in the glass and in the air on Alice's side. Pay attention to their spacing (at least qualitatively). For example, should the crest lines have the same spacing in all three regions? Explain your response.

6 The photon

There is something fascinating about science. One gets such wholesale returns of conjecture out of such a trifling investment of fact.

Mark Twain,
Life on the Mississippi

6.1 Preview

The last three chapters developed a wave theory of light. Indeed, they suggested that light is an electromagnetic wave, and then they marshaled experimental support for that view. The wave theory of light is correct – but it is not the whole story. Some other idea complements that theory. In this chapter, new experiments compel us to develop the alternative idea, and in chapter 7 we learn how the wave theory and the new idea complement each other.

6.2 The photoelectric effect

In this day and age, we are accustomed to the idea that sunlight is a source of energy. Let us investigate the connection between light (in general) and energy.

If you are not thoroughly familiar with how the word "energy" is used in physics, now is the right time to read appendix A, Energy. Knowing what physicists mean with their words will help you to understand the line of reasoning.

Electron escape

When an electron absorbs light, it acquires additional energy. If the electron is in a piece of metal, it may be able to escape from the metal. The positive charges in the metal tend to hold the electrons back and prevent their escape. (To be sure, the electrons can move about quite freely *inside* a good electrical conductor.) If, however, an electron gets a lot of extra

141

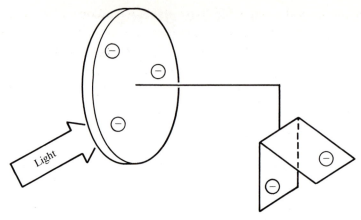

Figure 6.1 Apparatus for studying the photoelectric effect. A zinc disk is seen almost edge-on. Its diameter is about the length of your hand. Connected to the disk by the horizontal wire is the electroscope: a rigid metal strip and a flexible metal foil. Excess electrons, symbolized by circles with a minus sign inside, are on *both* sides of the zinc disk as well as on the rigid strip and the foil.

energy, it can escape the clutches of the attractive force exerted by the positive charges. This paragraph sets the background, and we can turn to new experiments.

Figure 6.1 displays the apparatus for our investigation. The equipment consists of a disk made of the metal zinc and then a device for assessing how much negative charge is on the disk. A wire runs horizontally from the disk to a vertical strip of rigid metal and a very light, flexible metal foil. When the disk is charged up with many excess electrons, their mutual electric repulsion pushes some of the excess electrons out through the wire and to the strip and foil. The excess electrons on the vertical strip repel the excess electrons on the light foil and lift the foil in defiance of gravity. The more excess charge, the more repulsion, and the farther the foil is pushed away and upward. The device is called an *electroscope* and appropriately so, for it enables us to "see electricity" or at least to discern the presence of excess electrons.

We charge up the disk (and electroscope) with excess electrons; the foil flips upward, showing plenty of electrons to be present. Then we illuminate the zinc disk with light from a 100-watt incandescent light bulb. The disk warms up as it absorbs some of the light, but no other effect is noticeable. The foil stays put, indicating that no electrons are escaping from the zinc.

Next, we try a mercury vapor lamp. The hot gaseous mercury provides a greenish-white light (the visible part) and also ultraviolet light. (A mercury vapor lamp finds a role as a germicidal lamp: the ultraviolet kills bacteria and other one-celled organisms.) Now the electroscope's foil begins to fall. The longer we illuminate the disk, the more the foil descends, and soon the lamp has totally discharged the disk: all the excess electrons have escaped.

Was the ultraviolet light crucial? To test, we charge up the disk again. Between the lamp and the disk, we interpose a wafer of the mineral mica; it absorbs ultraviolet but transmits visible light. Now the mercury vapor lamp has no effect: the foil does not budge. But pull away the mica, and the foil promptly drops. The ultraviolet light is crucial.

These few experiments are both satisfying and puzzling. Visible light – even lots of it – does not enable the electrons to escape. Ultraviolet light does.

The two kinds of light – visible and ultraviolet – are both forms of electromagnetic waves; intrinsically, they differ only in the values of their wavelengths and frequencies. Our experiments tell us that some relationship exists between the wavelength λ (or frequency f) of the light and the energy that the electrons can pick up from the light.

What we have found here qualitatively can be made quantitative, as follows.

When ultraviolet light of a specific frequency, such as $f=2\times10^{15}$ oscillations/second, is shone on metallic zinc, electrons escape. Some of those electrons have higher speed and hence have more kinetic energy than other electrons that have also escaped; there is a range of kinetic energies, from essentially zero up to some maximum. That maximum we describe by the phrase "maximum energy of electron after having escaped." (Lest there be a misunderstanding, note that the phrase does *not* mean that an electron's energy varies after the electron escapes and that we are looking at the energy's maximum value over time. Rather, we examine all the electrons that escape and find the maximum among their energies.) Figure 6.2 plots the "maximum energy of electron after having escaped" against the frequency of the light for two metals: zinc and lithium.

The graph for zinc starts in the ultraviolet and runs diagonally up and to the right. The lowest frequency that enables electrons to escape is called the *threshold frequency*. Already our experiments tell us that zinc has its threshold beyond the visible and, indeed, somewhere in the ultraviolet. For metallic lithium, the graph is similar, but the threshold lies in the visible, at green light.

The experiments that produced figure 6.2 lead to three generalizations:

(1) The higher the frequency, the larger the maximum electron energy after escape, and the relationship is a straight-line relationship.

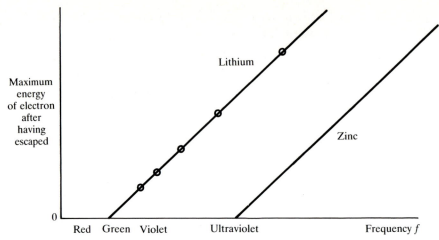

Figure 6.2 For electrons that escape from a metal, here is the experimental relationship between the frequency of the light and the energy of the electrons. The vertical axis starts at zero energy. The frequency scale starts near the frequency of red light, not at zero frequency. The tiny circles along the lithium graph represent some of Robert Millikan's data, discussed in section 6.4.

(2) The slope of the graph is the same for all metals; only the threshold varies from one metal to another.

(3) Neither threshold nor slope depends on the intensity of the light. (The figure itself does not exhibit this fact; it is added information. The intensity does, of course, have an influence on the escape process. The number of electrons ejected per second is proportional to the light intensity, provided any electrons escape at all.)

For a specific metal, we can organize the information presented by the figure in a tidy mathematical form. Because the maximum electron energy grows linearly with the frequency f, we can adapt the expression for a straight-line graph, "$y=mx+b$," to write

$$\begin{pmatrix} \text{Maximum energy of} \\ \text{electron after escape} \end{pmatrix} = \begin{pmatrix} \text{slope of} \\ \text{graph line} \end{pmatrix} f + \begin{pmatrix} \text{a constant that depends} \\ \text{on the material} \end{pmatrix}. \quad (1)$$

This equation applies for frequencies at threshold and above.

The slope is the same for all metals; we denote it by the letter h (for historical reasons, to be described later). At threshold, the "maximum energy of electron after escape" is zero. That condition enables us to express the constant in equation (1) in terms of h and the frequency at threshold, $f_{\text{threshold}}$, as follows:

$$\left(\begin{array}{c}\text{Maximum energy of}\\\text{electron after escape}\end{array}\right) = hf - hf_{\text{threshold}}. \tag{2}$$

This equation says that "maximum energy of electron after escape" is zero when f equals $f_{\text{threshold}}$ and rises linearly when f exceeds the threshold. In equation (2), the slope constant h is the same for all metals, but the threshold frequency depends on the specific metal, being in the ultraviolet for zinc but in the green for lithium.

A pause for a name is in order. The phenomenon at hand – that light can eject electrons from a metal and that the process has a threshold – is called the *photoelectric effect*. The composition is apt, for the word "photo" comes from the Greek root *photos*, meaning "light." We have an effect of light on electrons, ejecting them from their original location. Heinrich Hertz discovered the photoelectric effect when ultraviolet light inadvertently struck the metal of his detector for electromagnetic waves. The delightful irony of this will emerge later.

Now we return to question (2). The law of conservation of energy, applied to the escaping electron, also makes a statement about "maximum energy of electron after escape." We will work out the statement and compare it with equation (2).

The energy budget runs like this. The electron has some "initial energy" while it is in the metal and before it has absorbed any light. That is the starting value of the electron's energy. Then the electron acquires "energy absorbed from light." The electron is now able to escape from the metal despite the attractive forces exerted by the positive charges in the metal. But some energy is expended as the electron flees those attractive forces; so we subtract "energy needed to escape." What is left is the energy that the electron can call its own outside the metal. Among all electrons outside, the maximum such value is what we have called "maximum energy of electron after escape." Succinctly put, and in an advantageous order, the energy budget is this:

$$\left(\begin{array}{c}\text{Maximum energy of}\\\text{electron after escape}\end{array}\right) = \left(\begin{array}{c}\text{energy absorbed}\\\text{from light}\end{array}\right) + \left(\begin{array}{c}\text{initial}\\\text{energy}\end{array}\right) - \left(\begin{array}{c}\text{energy needed}\\\text{to escape}\end{array}\right). \tag{3}$$

On the right-hand side, the first term *could* depend solely on the light itself. The combination of the second and third terms may surely vary from one material to another. Moreover, that combination must have a net negative value, or else electrons would escape spontaneously, which is not so. The combination could determine the threshold for escape of electrons.

Equations (2) and (3) have the same quantity on the left-hand side. When we compare their right-hand sides, we see that the first term in one equation *could* correspond perfectly to the first term in the other. The comparison leads us to conjecture that

$$\left(\begin{array}{c}\text{Energy absorbed by electron} \\ \text{when light is absorbed}\end{array}\right) = hf. \tag{4}$$

The import is this: the amount of energy absorbed is never $0.2hf$ or $1.7hf$ or πhf or .... Rather, the exchange of energy between the light beam and the electrons is in "lumps," each of amount hf.

Gas station analogy

An analogy may be helpful. When you drive into a gas station, you may pump any amount of gas you want: 1 gallon or 8 gallons or 0.2 gallons or 1.7 gallons or even π gallons. Oil, however, comes in quarts only. You may buy 1 quart or 2 quarts or 3 quarts or any integral number of quarts. The exchange of oil between the gas station and you is *quantized*. No, not "quartized," but "quantized," from the Latin root

> *quantum*: neuter of the Latin "quantus," meaning "how much."

Wordplay aside, the meaning of "quantized" is that the exchange occurs in *indivisible amounts* of some specific size.

What appeared first in our development as a slope constant – the constant h – appeared in a related context when the German physicist Max Planck discovered quantum theory in 1900. Planck was studying theoretically the radiation in a hot oven and its interaction with the metal walls. The radiation has energy (the energy of electromagnetic waves), and the radiation may exchange energy with the atoms that form the oven walls. Suffice it to say that Planck selected the letter h for a new constant in his theory, a constant that related a frequency to an energy. Experiment required the constant to have the numerical value 6.6×10^{-34} joule seconds. Those are units of energy×seconds, just right for equation (4). Because a frequency has units of oscillations per second, the product hf has the units of (energy×seconds)×(oscillations per second); cancellation of "seconds" yields the units of energy (for "oscillations" do not count as a unit). The physics community subsequently named the constant h in Planck's honor, and so, for us, h is *Planck's constant*.

The photon enters

Equation (4) describes algebraically the exchange of energy between light and an electron. We can venture another step, from an electron-plus-light property to a light property alone. Here are two conjectures.

(a) The energy associated with the light beam (considered by itself) comes in lumps.

(b) The amount of energy associated with a lump is always hf and so depends on the color (or frequency f) of the light.

If these conjectures are correct, they introduce a kind of *graininess* into the theory of light.

In turn, such graininess introduces the *photon* (again from the Greek root *photos*, light). A photon is a particle of light. If a light beam has a frequency f, then a photon of that light has an amount of energy equal to hf, where h is Planck's constant:

$$\text{Energy of photon} = hf. \qquad (5)$$

The idea of such a particle of light was suggested by Albert Einstein in 1905. Section 6.4 will say more about the history of that suggestion. In 1926, the American chemist Gilbert Lewis coined the word "photon," and it has become the common name for Einstein's particle of light. The "-on" ending continues the tradition set by "electron" and "proton," electrically charged particles of matter, and is indeed merely a Greek neuter ending for nouns.

Right now we need to be clear about our conceptual framework. In chapter 2, we built a particle theory of light – Newton's particle theory – only to find it destroyed by experiment, notably by Foucault's experiment on the speed of light in water. We have also built a very successful wave theory of light, one based on electromagnetic waves. In a pure wave view, energy would be delivered to electrons in a metal continuously, like gasoline pumped into your car's gas tank. The notion of exchanging energy in indivisible amounts is foreign to a wave theory, where things happen smoothly. So let us be hesitant and skeptical about accepting a new particle theory of light.

Specifically, we must be careful to distinguish between

(a) graininess in the *interaction* of light and electric charges and

(b) graininess in light *per se*.

To be sure, by "a photon," physicists do mean "a particle of light," but we will have to see how tenable that concept is or how circumspectly it must be used. We are entering a domain where physics is subtle and where fine distinctions must be made. There is no straight-forward route, no simple picture. Rather, we face the challenge and delight of an intricate design, one of nature's most subtle. But let us go on.

6.3 The Compton effect

X-rays

Wilhelm Roentgen was professor of physics at the University of Würzburg, Germany, when he discovered X-rays in 1895. The discovery was entirely serendipitous; Roentgen was merely studying a beam of electrons in a highly-evacuated glass vessel. When the electrons, moving at great speed, slammed into the glass wall, they produced a very penetrating kind of radiation – a wholly unexpected occurrence. Roentgen first noticed the radiation when it caused a paper coated with barium platinocyanide to glow. The chemical compound was a standard detector of ultraviolet light, which causes the chemical to fluoresce, that is, to emit visible light after it has absorbed ultraviolet light. But Roentgen's evacuated vessel was tightly covered with black cardboard, and so no ultraviolet light could emerge from it. The glow must be produced by some other kind of radiation. Wood, leather, cardboard, and human flesh were easily penetrated. A book of a thousand pages was as nothing to the rays. Metal, however, absorbed the rays strongly. And the radiation blackened photographic film; it "exposed" the film.

Indeed, the radiation produced distinct shadows on photographic film. Roentgen became quite fond of them. When he announced the discovery of the new radiation, Roentgen wrote, "I possess, for instance, photographs of . . . the shadow of the bones of the hand, the shadow of a covered wire wrapped on a wooden spool, of a set of weights enclosed in a box . . . ." Earlier in the paper, he noted that "the darker shadow of the bones is seen within the slightly dark shadow image of the hand itself." The medical profession seized on Roentgen's discovery eagerly, and the new radiation quickly became a diagnostic tool in hospitals all over the world. Roentgen, however, could not determine what the "rays" were made of, and so he called them "X-rays": unknown rays.

Our own experience with visible light suggests what one might look for in trying to determine the intrinsic nature of X-rays. Reflection and refraction come to mind. Roentgen looked for both – and found neither. The experiments were difficult; the failure to find reflection and refraction might just mean one had not tried the right surfaces, prisms, and angles.

Interference was crucial in supporting the wave theory of visible light. Would X-rays show interference? In 1912, Max von Laue and two colleagues demonstrated that the answer was a resounding "yes." Simultaneously, they measured the wavelength. Since that time, there has been no doubt that X-rays are electromagnetic waves whose wavelengths are much shorter than those of visible light. (X-rays appeared beyond the ultraviolet on the electromagnetic spectrum in figure 5.13. And you may be relieved to know that X-rays were later found to reflect from surfaces and to be refracted by many materials.)

Compton

Among the physicists studying X-rays in the second decade of the twentieth century was Arthur Holly Compton. He received his Ph.D. from Princeton University in 1916, and two years later he turned his attention to the "scattering" of X-rays. Figure 6.3 illustrates the context. A beam of X-rays impinges on a block of graphite (which is carbon in the form of many tiny crystals). Most of the beam is transmitted through the block, but some X-rays emerge at 90 degrees to the original beam or at some other angle. One says that some of the X-rays have been "scattered" off to the side.

For over a decade, the scattered X-rays had been known to have less penetrating ability than the original beam, but how they were produced remained unknown. Compton measured the wavelength of the X-rays scattered at 90 degrees to the original direction; he found the wavelength to be a bit longer than the wavelength of the X-rays in the original beam (about 2×10^{-12} meters longer, certainly a tiny amount but measurable). Earlier, Compton had studied the scattering of gamma rays, a cousin of X-rays in

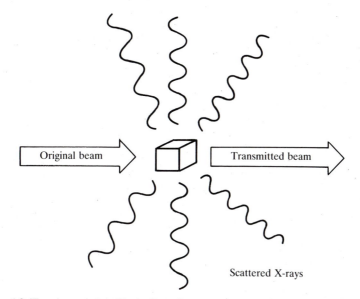

Original beam

Transmitted beam

Scattered X-rays

Figure 6.3 The electrons in a block of graphite scatter some of the X-rays in the original beam.

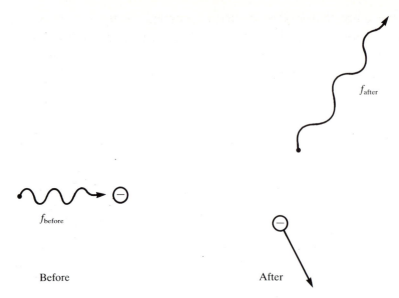

Figure 6.4 The photon idea applied to the scattering of X-rays. "Before" and "after" depict the situation before and after interaction between the X-rays and the electron. Each dot-and-wiggly-arrow represents a photon. The frequencies of the associated electromagnetic waves are denoted by f_{before} and f_{after}. The circle with a negative sign inside represents the electron; an attached arrow represents the electron's velocity. Bear in mind that these sketches are *representations* only; the sketches do not purport to show what a photon or an electron "really" looks like.

the electromagnetic spectrum. He had found that the wavelength of scattered gamma rays increased more and more as he examined rays scattered through greater angles from the straight-ahead direction. A pure electromagnetic wave theory had no plausible explanation for this behavior of X-rays and gamma rays, and so here was a puzzle indeed.

Compton wrestled with the conundrum for several years and found himself driven farther from a wave description and more toward a photon description. Ultimately, he concluded that he must accept the photon idea in its own right and apply it here in all possible detail. The change of heart produced a brilliant solution to the puzzle. Let's see how it goes.

Figure 6.4 sketches the context when we adopt a photon picture of what is going on. The graphite block is reduced to its barest essential: an electron initially at rest. (The carbon nuclei are too massive to be significant here. Although the electrons in a carbon atom are in motion, their speeds are sufficiently low that we may safely ignore the motion here, and we may likewise ignore the electrical forces that attract the electrons to the carbon nuclei. Without judicious idealization, physics would founder in a sea of minutiae.) The beam of X-rays is represented – at least as the interaction with the graphite block is about to commence – by a photon of energy hf_{before}, where f_{before} is the frequency of the original beam (thought of as an electromagnetic wave).

The interaction is similar to a collision between two billiard balls, one initially at rest, the other rolled along the green felt. The original photon is absorbed; another photon is emitted; and the electron recoils.

We can apply here the law of conservation of energy. The initial kinetic energy of the electron is zero (because the electron is at rest). Hence, when we equate total initial energy to total final energy, we have

$$hf_{before}+zero=hf_{after}+(\text{kinetic energy of recoiling electron}).$$

Because the recoiling electron's kinetic energy is a positive quantity, the product hf_{after} must be less than the product hf_{before}. Thus f_{after} is less than f_{before}. Because the product of frequency and wavelength equals c (in vacuum), a diminution in the frequency is accompanied by an increase in the wavelength. Such an increase is precisely what Compton had found.

Momentum for a photon

Compton went further. When a tennis ball hurtles across the net, we say it has a lot of momentum. Within physics, the precise definition of *momentum* is by a product: momentum equals mass times velocity. In turn, the word *mass* means "inertia," "sluggishness," or "the inherent reluctance to undergo a change in velocity." (To get a real "feel" for inertia, you can shake a lead brick. Once you have gotten the brick moving away from you and then try to haul it back, the brick tugs on your arms. You feel its inertia, its reluctance to undergo a change in motion.) Returning now to momentum, we note that a moving electron has momentum (by virtue of its inertia and its velocity).

In an isolated physical system such as our electron and the original photon, the total amount of momentum does not change with time. How can this principle – the conservation of momentum – be upheld here? After all, initially the electron has no momentum; when recoiling, it has some. The direct way out is to ascribe momentum to a photon (as well as energy). Then the total momentum at the beginning is not zero because the incoming photon has momentum. Moreover, the scattered photon can contribute to the total momentum in the final situation. Conservation of total momentum remains a tenable proposition.

How do we express the momentum of a photon? That is, what is the expression analogous to the relation

$$\text{Energy of photon} = hf?$$

We can build a line of reasoning based on the units in which quantities are expressed, such as meters per second for velocity.

First, we ask, on what quantities could a photon's momentum plausibly depend? The momentum could depend on the frequency (as does the photon's energy). It could depend on the speed of light c. (Once we have said this much, there is no need to include explicitly the wavelength, for it can be written in terms of f and c, to wit, $\lambda = c/f$.) Because the very idea of a photon is wrapped up with Planck's constant, we surely expect the momentum to depend on Planck's constant h. And that exhausts the supply of potential candidates.

Next, we ask, how can we combine f, c, and h so that they form an expression with the units of momentum? As mass times velocity, momentum has the units of kilograms times (meters per second). We need to construct a combination with those units. The product hf is a good start, for we know it has the units of energy. Appendix A, on energy, establishes that energy's units are kilograms times the square of (meters per second). If we divide hf by c, which has the units of meters per second, we have a combination with the units of momentum. So we can say

$$\text{(Momentum of photon) is proportional to } \frac{hf}{c}.$$

No other combination of our three ingredients will yield the correct units.

Now we face the issue of the proportionality constant. Our analysis with "units" can say nothing about that constant. More, however, was available to Compton and to the physicists of his era. Already James Clerk Maxwell had established that a burst of electromagnetic radiation carries not only energy but also momentum. For example, the radio signal en route to a planetary probe (which we discussed in section 5.1) carries momentum as well as energy across otherwise-empty space. The momentum manifests itself when an elctromagnetic wave is absorbed. The electric and magnetic fields exert forces on the electric charges in the absorbing material and thereby impart momentum to those charges. Maxwell was able to calculate the relationship of momentum (in magnitude) to energy; he found that the momentum in a burst of radiation was exactly $(1/c)$ times the energy in the burst.

Maxwell's work was done decades before Planck's constant emerged on the physics scene. Our result, arrived at by reasoning with units, has indeed the form "energy/c," in agreement with Maxwell's classical result. To ensure full consistency, we must take the proportionality constant to be precisely 1. Thus we arrive at our goal:

$$\text{Momentum of photon} = \frac{hf}{c}.$$

The combination of energy conservation and momentum conservation enabled Compton to analyze the scattering of X-rays in quantitative detail. He could make several predictions, the most obvious being that the scattering of X-rays should be accompanied by the recoiling of electrons. Such electrons had not been seen – because no one had looked for them. Within a few months, they were found – by C. T. R. Wilson in England and by Walter Bothe in Germany.

Compton could compute how the wavelength of the scattered X-rays should increase as he examined radiation at greater angles from the straight-ahead direction. He ran a new set of experiments, measuring systematically the wavelength of X-rays as a function of the angle at which they emerge from the graphite block. In April of 1923, Compton reported to the American Physical Society that this test of the theory worked out excellently. He had, in fact, been cautious lest his enthusiasm for his theory delude him:

> I was afraid of being influenced by finding what I was looking for,
> so I got to help me in the laboratory an assistant who did not know
> at all what I had in mind. I made the spectrometer settings while
> he took the readings. Not knowing what we were looking for, he
> felt that the readings were very erratic. After the experiment was
> over, he remarked to me: "It was too bad, wasn't it, Professor,
> that the apparatus wasn't working so well today?" He was distur-
> bed by the fact that the readings went up and down, and he had no
> idea that they were just the kind of thing I wanted.

Finally, one could calculate that, if the electron recoiled at angle so and so, then the scattered X-ray must go off at angle thus and such. The experimental test was especially difficult. The track of the recoiling electron had to be made visible, and the scattered X-ray had to be detected. By 1925, Compton and a colleague succeeded in the task, and the theory's angular relationship between recoil electron and scattered X-ray was confirmed, too.

The entire process – the sequence shown in figure 6.4 – now carries the name *the Compton effect*. Indeed, the phrase is taken to convey the physical process so fully that, when Compton shared a Nobel Prize in 1927, the citation read "for his discovery of the effect named after him."

For us, the Compton effect has two implications. First, it provides striking support for the photon picture of light. Second, it introduces us to the idea of ascribing momentum to a photon, and the analysis gives us an explicit expression for the magnitude of that momentum. (The direction of the momentum is simply the direction in which the light is traveling.)

Of course, you may marvel that we got to the expression hf/c for a photon's momentum when we started with the definition of momentum as the product of mass times velocity. The concept of momentum entered physics in the 1600s and was applied to commonplace objects: apples, cannon balls, and planets. Physics found momentum to be a useful notion and a quantity that is conserved in a physical system which is isolated from outside influences. Physics then extended the concept to other "objects," such as photons, relying on indirect reasoning (as we did) and tailoring the extended definition so that momentum – now generalized – would be conserved in an isolated system. (Specifically, Maxwell imposed conservation of momentum when he worked out the momentum carried by an electromagnetic wave.) That process of judicious enlargement is how physics operates when it extends the purview of a fruitful idea.

Having noted these points, we can turn to another aspect of the photon's history.

6.4 Einstein's Nobel Prize

In 1922, Albert Einstein received the Nobel Prize for Physics. The citation read "for his services to Theoretical Physics and especially for his discovery of the law of the photoelectric effect." The citation is really quite curious, and several stories lie behind it.

In chapter 9, we will begin our study of Einstein's special theory of relativity. Einstein published that theory in 1905. He spent much of the following ten years in broadening the scope of the theory, and in 1916 he published his general theory of relativity. It is a combination of the special theory and a theory of gravity. By 1922, experimental support for both theories – the special and the general – had become substantial. The support was not conclusive, to be sure, but some long-standing difficulties in physics had been cleared up, and at least one striking prediction – that the path of starlight is bent by the sun's gravity – had been dramatically confirmed. Why does the citation fail to mention the theories of relativity?

The year 1905 had been an extraordinary year for Einstein. Still a young man of twenty-six years, he published not only his special theory of relativity but also a seminal paper on molecular motion and the paper in which he suggested a particulate description of light, what would later be called the photon description of light.

The experimental basis for Einstein's particle-like theory of light was meager in the extreme. Heinrich Hertz, we noted, discovered the photoelectric effect while investigating electromagnetic waves. Sparks between two pieces of metal were essential elements of Hertz's apparatus, both in the transmitter circuitry and in the detector. In the course of his work, Hertz noticed that the "light" from one spark influenced the formation of another, distant spark. The other spark became easier to start and could be made to jump a larger gap between the two pieces of metal.

His curiosity piqued, Hertz dropped his work on electromagnetic waves for six months while he investigated this new phenomenon. In a series of conclusive tests, he found that visible light was not the source of the effect; rather, the cause was ultraviolet light from the first spark. In some way, the presence of ultraviolet light made it easier for a spark to form between two pieces of metal, one negatively charged, the other positively charged.

Having established this, Hertz returned to his electromagnetic waves. "As soon as I knew for certain that I was only dealing with an effect of ultra-violet light," Hertz wrote in 1892, "I put aside this investigation so as to direct my attention once more to the main question." He did not offer a theory of the photoelectric effect, and it remained for other physicists, notably Wilhelm Hallwachs and Philipp Lenard, to uncover further properties of the effect.

Despite their efforts, the data available in 1905, when Einstein thought about the effect, were scarcely better than what we had in section 6.2 after working with the zinc disk but *before* referring to figure 6.2. Nice straight-line graphs did not emerge until Robert Millikan, an American physicist, undertook a systematic study in the decade 1905 to 1915. The information listed as item (3) after we turned to figure 6.2 – the item concerning the influence of the intensity of the light – was well established by 1905, but little was known about how the "maximum energy of electron after escape" changes when the frequency of the light is varied.

Einstein begins his "photon" paper in a characteristically direct, iconoclastic fashion. The ways in which we think about gases and solid objects on the one hand and about light on the other hand, says Einstein, are strikingly different. The former we describe as particles; the latter, as waves. To be sure, concedes Einstein, the great success of Maxwell's electromagnetic wave theory of light is undeniable, and that theory is here to stay – at least for some purposes. But perhaps we can understand some other aspects of light – such as the photoelectric effect – much better if we describe light as a particle of some kind.

Suppose, suggests Einstein, that the energy of light comes in lumps. What theoretical consequences would this imply? Then he is off and running. Audacious and – in retrospect – brilliant.

Among the consequences which Einstein deduced was that the energy of a "lump" of radiation should be proportional to the frequency of the light. With theoretical consequence at hand, Einstein could *predict* that the graphs in figure 6.2 would be straight lines. A decade later, Millikan found them to be so. Moreover, Einstein could predict that the slope constant would be the same constant that Planck had introduced into physics in 1900 (and designated h). Millikan found that to be true also.

Here was substantial support for Einstein's ideas. How would the persons responsible for awarding Nobel prizes look upon it? Before we can suggest a plausible answer, we need to know a little about the purpose of the prizes themselves. And that, in turn, entails a look at the founder.

Alfred Nobel

As a young man, Alfred Nobel enjoyed writing poetry and was good at it. He was particularly fond of Percy Bysshe Shelley. In character, Nobel was a lonely, sensitive man, both in his youth and in his mature years. Though he never had the benefit of a university education, he became fluent in English, German, French, and Russian, four additions to his native Swedish. And he acquired a wide-ranging knowledge of literature.

Nobel's interests attracted him also to chemistry in general and to explosives in particular. Nitroglycerine fascinated him, and with it he invented dynamite, receiving a patent for that explosive in 1867. Nobel developed a financial network, spanning Europe and North America, for the manufacture of dynamite and other explosives of his invention. He poured skill and energy into industrial organization with as much success as he poured them into his chemistry.

A large fortune grew from the explosives business, and it was augmented by income from oil fields in Czarist Russia. Nonetheless, Nobel remained a retiring, considerate person, generous with his wealth. He was an idealist and a pacifist but skeptical about the prospects for disarmament. Certainly, he saw no easy route to the preservation of peace. Might there be a contradiction here: explosives manufacturer and pacifist? Perhaps, but Nobel's biographers have taken pains to point out that most of the explosives were designed and manufactured for civil blasting: the construction of highways, railroads, and canals, and also the driving of mine shafts. Nonetheless, the paradox is without a simple resolution.

Coupled with Nobel's concern about peace was his faith that natural science could improve the lot of humanity. A man of the latter half of the nineteenth century, Nobel lived in an era when technology and medicine made great advances, promising a better life – and delivering on the promise. (An acute awareness that technology carries environmental consequences lay 100 years in the future.)

When Nobel died in 1896, the public was surprised to learn that he left the greater part of his fortune to the world at large – in the sense that annual prizes were to be conferred on individuals who benefited humanity in any of five certain ways. Here is how Nobel's will stated the intention:

> The whole of my remaining realizable estate shall be dealt with in the following way:
>
> The capital shall be invested by my executors in safe securities and shall constitute a fund, the interest on which shall be annually distributed in the form of prizes to those who, during the preceding year, shall have conferred the greatest benefit on mankind. The said interest shall be divided into five equal parts, which shall be apportioned as follows: one part to the person who shall have made the most important discovery or invention within the field of physics; one part to the person who shall have made the most important chemical discovery or improvement; one part to the person who shall have made the most important discovery within the domain of physiology or medicine; one part to the person who shall have produced in the field of literature the most outstanding work of an idealistic tendency; and one part to the person who shall have done the most or the best work for fraternity among nations, for the abolition or reduction of standing armies and for the holding and promotion of peace congresses.

Physics, chemistry, physiology or medicine, literature, and peace: these are the five endeavors whose pursuit Nobel sought to reward and thereby to encourage. (That there is no Nobel prize in astronomy or mathematics, for example, is not to say that these subjects are any less valuable as intellectual studies. Rather, Nobel did not see them as benefiting humanity in the same measure as the sciences he selected.)

The first prizes were awarded in 1901, and the Swedish Academy of Sciences gave the physics prize to Wilhelm Roentgen for his discovery of X-rays. Roentgen was an admirable choice: X-rays were not only a remarkable scientific surprise, but in the few years since their discovery in 1895, their value in medicine had become overwhelmingly clear. To be sure, Roentgen had "conferred the greatest benefit on mankind" a little earlier than "during the preceding year," but the statutes had been written to allow for that: "the awards shall be made for the most recent achievements in the field of culture referred to in the will and for older works only if their significance has not become apparent until recently." The presentation speech could end proudly with the words, "Roentgen's discovery has already brought so much benefit to mankind that to reward it with the Nobel Prize fulfills the intention of the testator to a very high degree."

Back to Einstein

Now we can return to Einstein and his Nobel citation of 1922. Millikan had confirmed Einstein's predictions about the photoelectric effect; this greatly impressed the Nobel Committee for Physics of the Swedish Academy of Sciences. The presentation speech noted that "Einstein's law of the photo-electrical effect has been extremely rigorously tested by the American Mil-likan and his pupils and passed the test brilliantly." The Committee saw Einstein's work as the basis for photochemistry, the interaction of light with the atoms and molecules of chemistry. Here – perhaps – we see the benefit to humanity. The practical consequences of Einstein's photon theory flow – or, rather, in those years were often seen to flow – from equation (4) of section 6.2:

$$\left(\begin{array}{c} \text{Energy absorbed by electron} \\ \text{when light is absorbed} \end{array} \right) = hf.$$

This equation could be valid and fruitful regardless of whether photons exist or not, and that is how many physicists felt in 1922, Millikan included. The Committee could take a cautious stance and still award Einstein a Nobel prize, and that is what it seems to have done.

Although Einstein's theories of relativity are not mentioned explicitly in the citation, they do occupy the first paragraph of the presentation speech – but as theories that pertain to philosophy and whose validity is subject to active debate. Einstein's contribution to our understanding of molecular motion occupies the second paragraph. Then the speech turns to the photoelectric effect and devotes the bulk of the time to it. Annually, the Committee's task is to carry out the intent of Alfred Nobel's will. The members found a way to do that and yet to honor Einstein for his other "services to Theoretical Physics" as well.

6.5 The bare essentials

This chapter has more digressions than most, and so a capsule summary is in order.

The photoelectric effect told us that light delivers energy to electrons in a grainy fashion, in indivisible units of amount hf, where h is Planck's constant and f is the frequency of the light. This graininess in the *interaction* of light and electrons led us to conjecture that there is a graininess in *light itself*. Thus the photon enters. The photon is a *particle of light*, and about it we know two things:

$$\text{Energy of photon} = hf;$$

$$\text{Momentum of photon} = \frac{hf}{c}.$$

Arthur Compton analyzed the scattering of X-rays both experimentally and theoretically. His work provides strong evidence for the two relationships just cited – and so for the very existence of the photon.

But just how can we reconcile the photon – a *particle* of light – with the electromagnetic *wave* theory of light? That is the subject of the next chapter.

Additional resources

A twelve-page biographical sketch of Alfred Nobel appears in *Nobel: The Man and His Prizes*, third edition, edited by the Nobel Foundation and W. Odelberg (American Elsevier, New York, 1972). Particularly informative in the book is the piece by Ragnar Sohlman, who writes about Nobel's closing years and about the difficulties met in establishing the Nobel Prizes. A chemist, Sohlman was Nobel's personal assistant in the inventor's last three years of life and then became an executor of Nobel's will.

Abraham Pais addresses the subject "How Einstein Got the Nobel Prize" in his book *'Subtle is the Lord...,' The Science and the Life of Albert Einstein* (Oxford University Press, New York, 1982, pages 502–12). In 1922, Einstein actually received the Physics Nobel Prize for 1921. In some years, the Committee does not award the prize but rather defers it, and 1921 was one of those years. Niels Bohr received the Physics Prize for 1922 – in 1922.

Compton's recollections of the events surrounding his discovery are preserved in his article "The Scattering of X-rays as Particles," *American Journal of Physics*, volume 24, page 817 (1961), and reprinted in *Physics History from AAPT Journals* (American Association of Physics Teachers, College Park, MD, 1985). For further help in sorting fact from myth, I recommend Roger Stuewer's *The Compton Effect: Turning Point in Physics* (Science History Publications, New York, 1975).

"The evolution of the modern photon" is the topic pursued by Richard Kidd, James Ardini, and Anatol Anton in their article with that title: *American Journal of Physics*, volume 57, pages 27–35 (1989). The authors provide not only history but also food for thought as one grapples with the photon idea.

Questions

1. A certain metal surface ejects electrons when illuminated with green light but none when struck by yellow light. Do you expect that electrons will be able to escape when the surface is illuminated (*a*) by red light? (*b*) by blue light? Explain your reasoning.

2. A good mirror reflects about 80 per cent of the light energy that strikes it. How could you distinguish experimentally between the following two possibilities: (*a*) 20 per cent of the photons are not reflected; (*b*) all the photons are reflected but each with 20 per cent less energy?

3. If an electron in metallic cesium absorbs a photon of red light (specifically, $\lambda=6.6\times10^{-7}$ meters in vacuum), all the energy is used up in escaping from the attractive forces in the metal. There is no energy left over after the electron has gotten out (just barely).

Suppose blue-green light (with $\lambda=5\times10^{-7}$ meters) is used. What percentage of the photon's energy does the electron retain after it has escaped from the metal?

(Recall that a photon's energy is determined by the frequency of the light. You can, however, readily convert wavelength information into frequency information. If you do some algebra, you will find that Planck's constant cancels out in the final answer. That will save you some nasty multiplying – and you will see things more clearly.)

4. Some photographic materials can be handled safely in red light but are spoiled instantly when white light is turned on. How might you account for this?

5. More about the photoelectric effect. The text never addressed the question of why some electrons have less energy after escape than others. Suppose zinc is illuminated with ultraviolet light of a single frequency, for example, $f=1.5f_{\text{threshold}}$. Why might some electrons emerge with less energy than "maximum energy of electron after escape"?

7 The wave-particle duality

It is clear that the X-rays thus scattered proceed in direct quanta of radiant energy; in other words, that they act as photon particles. ... Time does not permit me to review the evidence that was accumulating in the meantime that gave full support likewise to the electromagnetic wave character of the X-rays. ... It became evident that though X-rays moved and did things as particles, they nevertheless have also the characteristic optical qualities that identify them as waves.

> A. H. Compton, 1961, recalling the events
> of the decade 1915–25.

7.1 Hertz revisited

In chapters 5 and 6, we noted that Heinrich Hertz discovered the photoelectric effect while investigating electromagnetic waves. A strong spark produces not only visible light but also ultraviolet light. When ultraviolet light from a spark in Hertz's transmitter circuitry struck the metal of his detector, the ultraviolet ejected some electrons. Those electrons, now present in the gap across which a spark was to jump, made it easier for a spark to jump. After all, a spark is an electric current, and the presence of the ejected electrons gave the spark a head start (for air is ordinarily a poor conductor of electricity, having few mobile charged particles). What Hertz perceived directly was that, when ultraviolet light shone on the detector, the spark was easier to start and could be made to jump a larger gap.

This was fascinating, and Hertz did spend six months establishing conclusively the essential role of ultraviolet light, but then he returned to electromagnetic waves. "As soon as I knew for certain that I was only dealing with an effect of ultra-violet light," Hertz wrote in 1892, "I put aside this investigation so as to direct my attention once more to the main question." Indeed, in the 1880s, the "main question" was whether electromagnetic waves existed; if so, whether they traveled at the speed of light; and, finally, whether visible light is an electromagnetic wave. Hertz provided the decisive experimental answer: yes.

And yet there is delightful irony here, for as Hertz was establishing the electromagnetic wave theory of light, he discovered the phenomenon – the photoelectric effect – that is most prominent in Einstein's invention of the photon theory of light. Of course, Hertz was not aware of what the photoelectric effect would lead to, and he did not even pause long enough to figure out how the ultraviolet light made the spark easier to form. Nonetheless, while Hertz was answering affirmatively the question, is light an electromagnetic wave?, he was also setting the stage – once again – for the question, is light a particle in character? Thus, in the fundamental research of a single person, we find the central question: is light a wave or a particle? That is the "main question" of this chapter.

7.2 A connection via probability

The photoelectric effect and the Compton effect demonstrate the graininess of light and led us to introduce the photon as a particle of light. Can we reconcile this graininess with the electromagnetic wave theory of light?

Experiment can guide us. We return to a double-slit interference experiment but specify very weak, dim light and arrange to register electronically the appearance of individual photons. Figure 7.1 sketches the apparatus with which such an experiment has actually been done. Light from the source goes through the pair of slits, passes through the transparent plate, and forms an interference pattern on the thin "photoelectric coating." The coating is a thin layer of special material in which the photoelectric effect works particularly well. When an electron in the coating absorbs a photon, the electron is ejected (toward the right). Fancy electronic circuitry collects the electron, determines the location where it was ejected from the coating, and plots a corresponding point on a monitor screen, a TV-like screen of especially high quality.

In the experiment, the light was so dim that only some 100 electrons were ejected in a second's time. The electronic circuitry could easily register them individually and plot their origin one by one. Here is how the authors describe the experiment:

> As measurement was started, bright dots appeared at random positions on the monitor screen. After 10 seconds had elapsed, the photon-counting image shown in [figure 7.2 (*a*)] had appeared. The image contains 1000 events, but the overall shape is not yet clearly defined. [Figure 7.2(*b*)] shows the image obtained after an

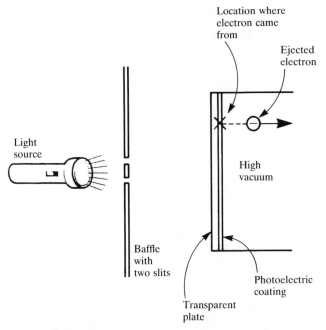

Figure 7.1 Apparatus for registering the appearance of individual photons in an interference pattern. We see the slits edge-on. The source produces ultraviolet light (having $\lambda = 2.5 \times 10^{-7}$ meters). The symbol $\times$ denotes a location in the interference pattern where a photon of that light has been absorbed and thus has enabled an electron to escape.

integration of 10 minutes. The total count accumulated by this time was 60,000, and the interference pattern is clearly seen.

When only 1000 photons have been registered, there is no full interference pattern. The pattern builds up slowly, and when very many photons have been registered – 60 000 here – the profile matches the intensity profile that we ordinarily see and that is predicted by the electromagnetic wave theory.

If one repeats the experiment, the locations where the first 1000 photons are registered are different, but the pattern that emerges in the long run is the same.

There seems to be no rhyme nor reason for where the first photon appears, *except* that it never appears where it should not (according to the electromagnetic wave theory) and it does appear more frequently where the pattern will later have many registered photons than where the pattern will have few.

Our inference is this: just where a photon appears is determined only

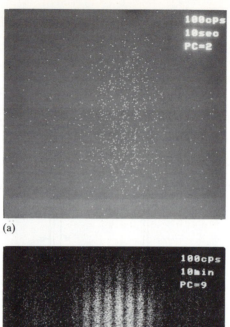

(a)

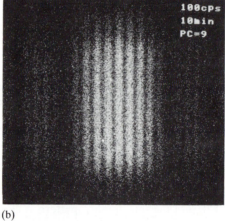

(b)

Figure 7.2 The appearance of individual photons in an interference pattern. (*a*) After 10 seconds have elapsed and 1000 photons have been registered. (*b*) After 10 minutes and 60 000 photons. The apparatus was developed by four Japanese physicists, T. Tsuchiya, E. Inuzuka, T. Kurono, and M. Hosoda; they reported their work in a paper entitled "Photon-counting imaging and its application," published in *Advances in Electronics and Electron Physics*, volume 64A, pages 21–31 (1985).

statistically. And so we need to talk about the *probability* of appearance in a specific small region.

The probability of appearance should be high if we predict, from wave theory, that the light intensity will be high in the specific region. The reasoning is direct: high light intensity means lots of energy being deposited and hence lots of photons appearing, each contributing energy hf.

For more detail, let us examine "light intensity" on some small patch of area from two points of view. In each case, the essential proportionality will be displayed first, and then the text will justify it. (The symbol $\propto$ stands for the phrase "is proportional to.")

Photon description

$$\begin{pmatrix} \text{Light} \\ \text{intensity} \end{pmatrix} \propto \begin{pmatrix} \text{rate at which photons} \\ \text{appear on patch} \end{pmatrix} \begin{pmatrix} \text{energy} \\ \text{per photon} \end{pmatrix} \propto \begin{pmatrix} \text{probability} \\ \text{of appearance} \end{pmatrix} hf.$$

Light intensity is proportional to the rate at which energy arrives on a small patch of area. That rate is given by the product of the rate at which photons appear on the patch times the energy that each photon delivers. In the long run, the rate at which photons appear is proportional to the probability of appearance, and so the second proportionality follows.

Electromagnetic wave description

$$\begin{pmatrix} \text{Light} \\ \text{intensity} \end{pmatrix} \propto \begin{pmatrix} \text{rate at which } \mathbf{E} \text{ of wave imparts} \\ \text{energy to charges in patch} \end{pmatrix} \propto \begin{pmatrix} \text{square of } \mathbf{E} \text{ of} \\ \text{electromagnetic wave} \end{pmatrix}.$$

In the electromagnetic wave picture, the electric field of the wave exerts a force on the electric charges in the patch of area. Those charges may be electrons in a metal or electrons in an atom or molecule; it does not matter. The electric force pushes the charges and imparts energy to them. That justifies the first proportionality.

For the second proportionality, we note that the larger the electric field, the larger the force it exerts on the charges and hence the more energy it imparts. The rate of energy transfer must grow with the size of the electric field, and it should not depend on the direction of the electric field: up versus down or left versus right. The square of the electric field meets both criteria and actually is the correct relationship. We cannot pretend to have derived it rigorously, but nothing essential in our subsequent reasoning depends on there being precisely a square in the relationship, and so we will not pause for a lengthy derivation.

We have now two expressions for the light intensity. We have reason to believe that each expression is correct – within its own description of light. Therefore the factors on the far right-hand sides of the two expressions must be proportional to each other. Once the frequency of the light source is set, the factor hf is fixed, and we may drop it from consideration. Thus we infer that

$$\begin{pmatrix} \text{probability of} \\ \text{appearance of photon} \end{pmatrix} \text{ is proportional to } \begin{pmatrix} \text{square of } \mathbf{E} \text{ of} \\ \text{electromagnetic wave} \end{pmatrix}. \quad (1)$$

The electromagnetic wave theory gives us a *probability profile* for observing photons, that is, for observing energy transfers in a grainy fashion. Here is a crucial link indeed.

An interference pattern consists of darks, maximum brights, and intermediate levels of light intensity. The wave theory calculates such an intensity profile by computing where the electric field **E** will be zero or maximum or some intermediate magnitude. Equation (1) tells us that these variations in the size of the electric field can be translated into a probability profile for the appearance of photons: zero probability at some locations, maximum probability at others, and intermediate values in between. This probabilistic link begins to reconcile the graininess of the photon aspect of light with the smoothness of the wave aspect.

7.3 A trajectory for a photon?

In some respects, a photon is like a tiny baseball. A photon does, after all, have energy and momentum. But does a photon have other properties in common with a baseball? Specifically, can we ascribe to a photon a *definite, continuous trajectory*, although perhaps only a trajectory chosen at random?

We need to be clear about the question. First, the verb "to ascribe" means "to attribute" or "to assign." It does *not* mean "to predict." The issue before us is not whether we can predict a trajectory for a photon. Rather, the issue is whether we may say that a photon *has* a definite, continuous trajectory. The question of prediction can arise only if one can first establish that there is something to predict.

Second, just what does the phrase "definite, continuous trajectory" mean? If you visualize a high fly ball rising from the bat at home plate and sailing in a grand arc to the left fielder's outstretched mitt, you are visualizing a definite, continuous trajectory for a baseball.

Here is another example. Your Aunt Charlotte in San Francisco is coming to visit you in New York. You check with a travel agent and find that the only flights still available are these two: (*a*) fly from San Francisco to Chicago, change planes, and then fly to New York or (*b*) fly from San Francisco to St. Louis, change planes, and then fly on to New York. Quickly, you call Aunt Charlotte and give her the information; she is to make reservations herself. You cannot predict which route your aunt will take. Nonetheless, when you meet Aunt Charlotte at the information desk in the New York airport, you are confident that she flew either via Chicago

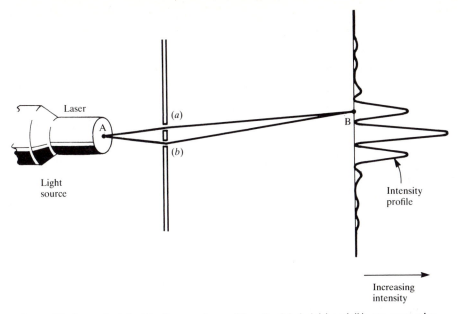

Figure 7.3 A standard double-slit experiment. The slits, labeled (*a*) and (*b*), are seen edge-on, and the usual cylindrical laser beam is wide enough to illuminate both slits. When both slits are open simultaneously, one gets the intensity profile shown on the right. Intensity is plotted rightward from the plane where photons appear. Shown also are two conceivable trajectories for a photon that leaves the laser light source at point A and appears at point B.

or via St. Louis. That is the "definiteness" of her trajectory. The "continuousness" lies in your conviction that your aunt was either on a plane or in an airport terminal at every moment throughout the trip; she did not just vanish in San Francisco and materialize in New York.

But enough of this prelude. What experiment can we do to answer the question, can we ascribe to a photon a definite, continuous trajectory?

Imagine a standard double-slit experiment but done with laser light of low intensity. Instead of viewing an interference pattern on a wall, we capture it on film in an otherwise-dark room, exposing the film for 10 seconds. Figure 7.3 shows the set-up and the ensuing intensity profile: the pattern of darks and brights. Despite the dimness of the light, very many photons contribute to exposing the film.

If a photon has a definite, continuous trajectory, then a photon that appears at point B went through either slit (*a*) or slit (*b*). (Remember your Aunt Charlotte here: she passed through either Chicago or St. Louis. Just which slit the photon went through may have been chosen in a random fashion; similarly, your aunt may have flipped a coin in choosing her route.)

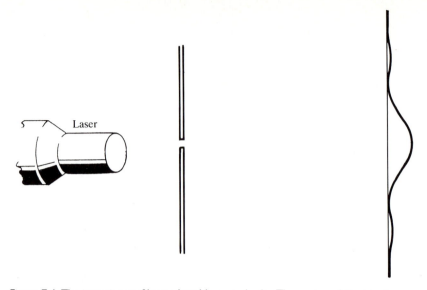

Figure 7.4 The intensity profile produced by a single slit. The center of the interference pattern lies directly opposite the single slit (because the laser beam runs perpendicular to the plane in which the slit is cut). The width of the slit here is the same as the width of each slit in figure 7.3. The scale for the intensity profile is the same in the present figure, in figure 7.3, and in the next figure.

If the first sentence of this paragraph is true, then we ought to get the same pattern on film if

(i) we block slit (*b*) and expose for 10 seconds and

(ii) then we block slit (*a*) and expose for 10 seconds.

Thus each slit is open for 10 seconds, as in the original situation. In short, if a photon has a definite, continuous trajectory and if many photons contribute to the interference pattern, we should get the same result if the slits are open *simultaneously* or *sequentially*.

What, in fact, is the final intensity profile if the slits are open sequentially? To answer this question, we need to know the intensity profile produced by a single slit whose width is equal to that of each slit in the double-slit pair. Section 4.7 developed the theory of the single slit, and experiment can provide the profile, too. Figure 7.4 shows the single-slit pattern that we need here.

The single slit produces a broad intensity profile, and we are to add two such profiles, one for each 10-second exposure. Figure 7.5 shows this done. The centers of the two profiles are to be separated by an amount equal to the center-to-center separation of the two slits in the double-slit situation.

Figure 7.5 The sum of two single-slit intensity profiles.

When the interference pattern is formed far from the slits, the broad profile is very much wider than the slit separation. Hence the result of adding the two profiles is very nearly the same as plopping one single-slit pattern down right on top of the other.

Now we are ready for the crucial comparison: the final intensity profile in figure 7.5 with the profile in figure 7.3. The sum of two separate single-slit patterns (slightly displaced) does not give the same intensity profile that a standard double-slit apparatus gives. The many valleys are missing. But if a photon has a definite, continuous trajectory, then it can make no difference whether the slits are open simultaneously or sequentially. Thus what would have to happen if a photon has a definite, continuous trajectory does not happen.

Our conclusion is profound: we cannot ascribe to a photon a definite, continuous trajectory, not even one chosen at random.

7.4 The wave-particle duality

In the last chapter, we moved from a pure wave theory of light to something with particle aspects. The photoelectric effect and the Compton effect compelled us to do so. Can we go all the way, to light as a hail of photons, with no wave aspects?

No. No pure hail-of-photons theory can explain or predict interference.

After all, when we visualize hail falling, we picture definite, continuous trajectories. The preceding section taught us, however, that we cannot ascribe such a trajectory to a photon. A "hail of photons" is not a tenable picture of how light behaves. Rather, double-slit interference mandates a wave-like character for light.

What's more, photons are very *unlike* tiny baseballs. To be sure, a photon has energy and momentum, but we cannot ascribe to a photon a definite, continuous trajectory. In an absolutely crucial way, a photon – whatever it is – differs from our commonsense conception of a "particle."

We are stuck. Neither a pure wave theory nor a pure particle (or photon) theory will work. Light is a more complicated phenomenon than any *single one* of our everyday concepts can describe.

About the best one can say is laid out in the following paragraphs.

If we want to know the overall pattern that light makes in space and time, we can use the electromagnetic wave theory. Generally this suffices for understanding interference, refraction, reflection, and phenomena like the effect of Polaroid sheets.

Neither the electromagnetic wave theory nor any other theory can tell us exactly where and when light energy will be transferred, in a grainy fashion, to individual charged particles (or be emitted by them). The electromagnetic wave theory does give us the probability of such an event. For the absorption of light, the wave theory provides the probability that a photon will appear (and then immediately disappear, as it is absorbed). Our prime example is the photoelectric effect; Compton's experiments go here, too. In such microscopic interactions of light and matter, the photon notion is dominant.

The two pictures – light as a wave and light as a particle – *complement* each other. Light is something with the *potentialities* for acting like a wave or a particle, two complementary aspects that are realized under different physical conditions. This complementarity is called the *wave-particle duality*.

We can capture the duality in an aphorism:

> light travels as a wave but departs and arrives as a particle.

The phenomena of interference, refraction, and reflection result when light travels from one place to another, and the wave theory provides an understanding of them. The emission and absorption of light depend on the interaction of light and electric charges (as in the photoelectric effect). Those processes occur when light departs from some location or arrives at one, and then the photon theory proves to be essential.

7.5 Understanding light: some further help

The previous section, entitled "The wave-particle duality," is the culmination of our study of light. Anything else is anti-climactic or ancillary. There are, however, a few more things worth saying before we go on and a few things to repeat, for you are now better prepared to place them in context. This section is *not* a summary of the first half of the book; that would be too ambitious a task. Rather, it is a collection of statements and comments designed to provide further help in understanding light.

- Light is a wave phenomenon (though this statement is not the whole story). Our best evidence for this proposition is interference, as in double-slit interference. Refraction (in its details) is very good evidence, too.

- Out of what is a light wave formed? Electric and magnetic fields.

- And what are electric and magnetic fields? They are defined operationally in terms of forces on charged particles or compass needles. They can exist in what is otherwise vacuum (as in the instance of a radio message sent to a planetary probe near Jupiter). The fields are fundamental: I cannot express them in terms of anything more primitive or more intuitive. We can visualize the pattern they make in space, but we cannot "see" them literally.

- A light wave is always formed from both electric and magnetic fields. The fields always point perpendicular to the propagation direction, that is, perpendicular to the direction in which the light is traveling.

- An electron is a tiny *electrically charged* particle. (The flow of electrons in a copper or aluminum wire is what carries "electricity" around your house.) An electron is quite distinct from a photon. For example, photons have no electric charge, and electrons can have any speed from zero up to the speed of light c (but not including that limiting value).

- When light interacts with electrons, the exchange of energy occurs in a grainy fashion. The amounts of energy exchanged is always hf, where h denotes Planck's constant and f is the frequency of the light wave. This graininess introduces the photon.

- In which way should I think of a photon as a particle? Indeed, as "a particle of light"?

(*a*) A photon has energy, and it has momentum. These are nice particle-like properties.

(*b*) We cannot ascribe to a photon a definite, continuous trajectory. That is distinctly *not* in accordance with our commonsense concept of a particle.

(*c*) It is tempting to think of a light beam as merely a hail of photons, but it is also incorrect. Don't do it. A wave traveling through space is a better picture.

(*d*) The graininess of light manifests itself primarily when light interacts with charged particles. Thus the photon notion is needed primarily at the start and finish of a light beam's travel through space. Indeed, the title of Einstein's "photon" paper was this: "On a heuristic point of view concerning the production and transformation of light." The word "heuristic" means "stimulating discovery or further investigation" and indicates a tentativeness. More significant for us is Einstein's specification of "production and transformation of light." We should understand "production" to mean the emission of light. By "transformation," Einstein meant the combination of absorption followed by emission. And in the body of the paper, Einstein devoted extensive space to the absorption of light by metals and molecules. Einstein had clearly in mind the interaction of light and electric charges when he introduced his particle theory of light.

- If you look back at figure 6.4, you will see a representation of a photon: a dot and wiggly line. It is a good representation, for it reminds us that the particle aspect of light – the dot – is intimately connected with the wave aspect – the undulating line. Though a good graphic representation, the dot and wiggly line are *not* a good model for a photon. Note especially that a photon does *not* behave like a tiny ball that bobs up and down during its flight. There is nothing in the laws of physics to suggest such behavior; rather, the laws forbid such behavior. Do not let an attractive representation mislead you into ascribing reality to it. A photon is no more a tiny bobbing sphere than the United States of America is an eagle with a bundle of arrows gripped by the talons of one foot and an olive branch by the talons of the other. Physics is not able to provide a valid, intuitive picture of a photon, and that is why none appears in this book. I would love to give you such a picture, but I cannot.

- Photons and waves are inextricably linked. Our major connections are the following three. Note that the photon idea appears on the left and the wave idea, on the right.

$$\text{Energy of photon} = hf;$$

$$\text{momentum of photon} = \frac{hf}{c};$$

$$\left(\begin{array}{c}\text{probability of}\\\text{appearance of photon}\end{array}\right) \text{ is proportional to } \left(\begin{array}{c}\text{square of } \mathbf{E} \text{ of}\\\text{electromagnetic wave}\end{array}\right).$$

- The question of when to use which concept – wave or photon – was answered (in part, anyway) in the preceding section. But to recapitulate briefly: use the wave picture for describing interference, refraction, reflection, and phenomena with Polaroid sheets; use the photon idea for describing the interaction of light and electric charge, that is, for describing absorption and emission of light.

- You may feel that understanding light is difficult. If so, then you are in good company; let me illustrate with a passage from a letter. Albert Einstein's best friend was Michele Besso. When they were in their 20s, the two of them were colleagues in the Patent Office in Bern, Switzerland, and they maintained a correspondence throughout the rest of their lives. In a letter dated 12 December 1951, when Einstein was 72 years old, Einstein wrote these lines to Besso:

> The entire 50 years of deliberate pondering have not brought me closer to an answer to the question "What are light quanta [that is, photons]?" Today, every Tom, Dick, and Harry believes that he knows, but he deceives himself.

  To be sure, Einstein always set high standards for what he would call "understanding;" yet his frustration points up the difficulty of the task.

- You may wonder whether, after all, physicists understand light. When it comes to equations, physicists understand light very well indeed. For example, theoretical calculations in the 1950s laid the foundation for the first lasers (which were constructed in the early 1960s). In no way were lasers accidental discoveries.

  Providing pictures to capture the essentials of what the equations have to say is quite a different task. Sometimes a wave picture will do the job; at other times, it will not. And the same is true for the photon description. But where one description fails, the other succeeds: therein lies the complementarity. And for us there is a lesson in humility: light is a more subtle phenomenon than any single one of our everyday concepts can describe.

Additional resources

Professor John King's classic 19-minute film, *Photons,* shows that a light beam delivers its energy in a fashion that is not only grainy but also random in time. As of this writing, the film was available for rent or purchase from Ward's Natural Science Establishment, P.O. Box 92912, 5100 West Henrietta Road, Rochester, New York 14692–9012.

Questions

1. A sentence or two will suffice for each part. Be cogent and terse.

(*a*) What property of the wave theory makes it unable to explain the photoelectric effect?

(*b*) Cite two respects in which a photon is quite *unlike* a little baseball.

(*c*) For which light phenomena do we need the wave theory?

(*d*) Write out three mathematical relationships that link the wave theory and the particle theory. (By "the particle theory," I mean the photon version, not Newton's version.) Be sure to point out where the wave theory makes its appearance and where the particle theory does. That is, on which side of the relationship (either equation or proportionality) does the wave theory appear and on which does the particle theory appear?

(*e*) How did we ultimately explain the "sidedness" of light?

2. A sentence or two will suffice for each part.

(*a*) Why did we abandon Newton's particle theory of light? (Not "why might we have," but "why did we?")

(*b*) For metallic lithium, the threshold for ejection of electrons by light lies at wavelength $\lambda=5.3\times10^{-7}$ meters, a green light. Specify a color of light that will eject electrons with energy to spare and another color that will certainly fail to eject electrons. Briefly explain why you chose the colors that you did.

(*c*) When a beam of yellow light passes from vacuum into water, which of the following properties does *not* change? Speed, frequency, or wavelength. Why?

(*d*) What experimental reason do we have to believe that a photon has momentum (as well as energy)?

(*e*) Why do we need to introduce "probability" into the description of light? Be very specific.

3. Can one prove a scientific theory? For example, can one prove Newton's particle theory of light? Or Maxwell's electromagnetic theory of light?

Before you set pen to paper, look up the verb "to prove" in a really good dictionary. (The *Oxford English Dictionary* would be optimum.) Be careful to distinguish two significantly different but relevant meanings for the verb. Delineate those two meanings as part of your response.

4. Suppose you set up a double-slit interference experiment with blue light. You use photographic film to register the bright and dark regions on the viewing surface. Fix your attention on the first dark region, where the photographic film is *not* exposed (because no blue light appears there). Now suppose you leave everything as it was but switch to an ultraviolet light source. For your earlier and present sources, the relationship of the wavelengths is this: $\lambda_{\text{ultraviolet}} = \frac{1}{2}\lambda_{\text{blue}}$. Will the film at the location "first dark for blue" remain unexposed? Explain your reasoning carefully.

8 Does the speed of light depend on the motion of the source of light?

It's hard to argue with a good experiment.

Phineas Oort

8.1 Transition

This chapter provides a transition: from our study of light for its own sake to our study of space and time as described by Albert Einstein's special theory of relativity. Our investigation of another property of light serves as the bridge. We ask the question, does the speed of light depend on the motion of the source of light? And we seek an experimental answer. The question sounds perfectly innocent – and in some ways it is, for no tricks are involved – but the consequences of the answer will be profound.

8.2 The question answered

Let us recall what *speed* means. Figure 8.1 shows two fixed locations, denoted A and B. For anything that travels in a straight line from location A to location B, what we mean by its speed is this:

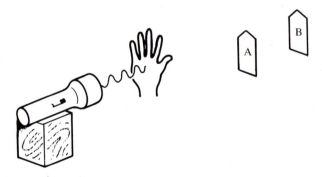

Figure 8.1 How to measure the speed of light (in principle!). The flashlight rests on a wooden block; A and B denote fixed locations along the path taken by a burst of light.

$$\text{Speed} = \frac{\text{distance traveled}}{\text{elapsed time}} = \frac{\text{separation of A and B}}{\text{time to get from A to B}}. \qquad (1)$$

To set the context further, we are going to imagine a couple of experiments before turning to the professionally-done experiments. So imagine a flashlight placed on a wooden block and aimed along the path from A to B. You can hold your hand in front of the flashlight to block the beam, then remove your hand, allow a burst of light to proceed down the path, and quickly replace your hand. To determine the speed of light, we need to note the time when the burst arrives at A, note when it arrives at B, and subtract to determine the "time to get from A to B." A meter stick will measure the separation of A and B. Then we have the numerator and denominator on the right-hand side in equation (1) and hence have the speed. Because the flashlight rested on the wooden block, these measurements give us the speed of light from a stationary source.

By 1964, careful professional measurement of the speed of light (in vacuum) from a stationary source gave the value

$$(\text{Speed of light})_{\text{source at rest}} = 2.997925 \times 10^8 \ \text{meters} / \text{second}. \qquad (2)$$

The uncertainty in this value was plus or minus 3 units in the last decimal place, that is, the correct last digit might be as small as $5-3=2$ or as large as $5+3=8$. You may notice that the round figure of 3×10^8 meters/second that we have used heretofore is quite close to the more accurate value.

But what if the light source moves? Does the speed of light change? And if so, by how much?

We can imagine taking the flashlight off the block, moving it uniformly toward location A, and repeating the previous experiment: use your hand as a shutter to allow a burst of light from the moving source to travel from A to B. This is fine in principle, but how would a professional do it?

If source motion has any influence on the speed of light, then surely the faster the source moves, the larger the influence will be. So, to maximize any possible influence, one should use a light source that moves very fast.

The 1964 experiment

We turn now to an experiment performed by Torsten Alväger and colleagues near Geneva, Switzerland, in 1964. As background, let me remind you of some of the particles of physics. We have talked extensively about electrons; they are negatively charged and have very little mass. The positive charge in an atomic nucleus is provided by one or more protons; they

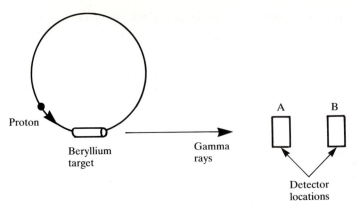

Figure 8.2 Measuring the speed of light from a rapidly-moving source of light. The neutral pion decays into two gamma rays while moving swiftly through the beryllium target. The detectors of gamma rays, labeled A and B, are 31 meters apart.

are roughly 2000 times as massive as an electron. A hydrogen atom consists of one proton and one electron; their mutual electrical attraction keeps them near each other, and thus they form an atom.

Electrons and protons are stable in both the colloquial and the technical sense. If left in isolation, an electron or proton remains just that; no spontaneous change occurs.

Other particles occur in nature, too, and now we need to learn about a new species: the pions. In mass, the pions are intermediate between the electron and the proton. Moreover, pions come in three sub-species: positively charged, negatively charged, and neutral. The pions are unstable. In particular, the neutral pion decays spontaneously into two gamma rays, that is, into two photons of very high energy. This is our first encounter with any kind of radioactivity; more will come later. Right now, the significance is this: by virtue of its decay into two gamma rays, a neutral pion is a source of light. And it can be a swiftly-moving source of light, as we shall see.

For us, the 1964 experiment starts with protons accelerated to high speed in a circular tunnel. Figure 8.2 shows their path. When the protons have been brought up to the desired speed, they collide with a "target:" a 2-centimeter length of wire made from metallic beryllium. A high-speed proton interacts with a proton in a beryllium nucleus; from the collision emerge the two protons and – sometimes – a neutral pion as well. The pion has been produced where none existed before. All three particles – the two protons and the pion – move on at high speed, but almost immediately the pion decays into two gamma rays. Thus the pion constitutes a light source moving at high speed.

The task now is to measure the speed of the gamma rays emitted in the decay of the pion. In effect, the physicists measure the following three quantities:

- time from production of light to arrival at A;
- time from production of light to arrival at B; and
- separation of A and B.

The difference of the two times is the travel time from A to B.
  Thus the physicists calculate

$$\left(\text{Speed of gamma rays from pion}\right) = \frac{\text{separation of A and B}}{t_{\text{production to B}} - t_{\text{production to A}}}$$

$$= 2.9979 \times 10^8 \;\; \text{meters}\,/\,\text{second}. \quad (3)$$

The value of "time" is represented by the letter t. The experimental uncertainty is plus or minus 4 units in the last decimal place.

When we compare the results in equations (2) and (3), we see that the speed of light from the moving source is the same as that from the stationary source. (That is, the two speeds agree to all the digits displayed in equation (3), and those are all the digits that the 1964 experiment can claim to have established. Only the last of those digits is subject to some uncertainty.)

But, you may well ask, how fast were the neutral pions going? Because they are uncharged, neutral pions are difficult to observe, and so their speed is difficult to measure. For charged pions produced in the same way, the speed has been measured directly: by measuring the time needed to travel a known distance. The speed of those pions is faster than 99 per cent of the speed of light itself (that is, the speed exceeds $0.99c$, where c is our familiar symbol for the speed of light in vacuum, 3×10^8 meters/second, in round numbers). So the light source in the 1964 experiment was moving faster than $0.99c$ and hence was moving rapidly indeed.

The inference

Our original question was this: does the speed of light depend on the motion of the source of light? On the basis of the 1964 experiment, we can answer "no" with confidence. Indeed, we elevate the question and answer to the status of a *principle* (as in "a fundamental truth").

> Principle: the motion of light is not affected by the motion of the source of light.

> Justification: generalization from the excellent 1964 experiment (and from other experiments not described here).

The principle stated here forms the second of two principles on which Einstein based the special theory of relativity. We will follow suit. But before we launch into relativity theory *per se*, there are a few items to attend to.

8.3 A few comments

Lest there be any misunderstanding, we are talking in this chapter – and in the rest of the book, for that matter – about the speed of light in vacuum. That is the speed denoted by the symbol c.

Next, you might wonder how the physicists can detect the same gamma ray twice, first when it passes location A and then when it arrives at location B. The response is that the physicists do not actually do that. Rather, here is how the timing proceeds. The protons in the circular ring are organized into bunches of protons with lots of space between adjacent bunches. (The bunches are spread out like several packs of runners on a circular track.) When a bunch of protons hits the target, it produces a burst of pions. The pions decay in flight and produce a burst of gamma rays. Some of those gamma rays are absorbed and detected at location A, and the remainder travel on to a backstop that absorbs them safely. This measurement gives, in effect, the "time from production to A." Then the detector is shifted farther from the target, to location B, and the procedure is repeated, yielding the "time from production to B." There are many particles – protons and pions – doing the same thing, and so it suffices to examine some of their gamma rays at A and others at B.

What may seem most bizarre to you are the pions themselves. They *are* intrinsically puzzling. Why do they have to enter our story? If we could readily get a flashlight to whip through the laboratory at 99 per cent of the speed of light, I would never have introduced the pions. Physicists would have done the experiment with moving flashlights, too, and we would have discussed that familiar light source. But objects the size of a flashlight cannot readily be accelerated to a speed approaching that of light. To find a really swiftly-moving source – swift even when compared with c – one needs to turn to objects of atomic or sub-atomic size.

That the pion decays into two gamma rays and thus disappears from the scene should not disturb us, either. Think of a fireworks display on the Fourth of July. A cardboard rocket, trailing sparks, ascends into the darkness and explodes in flight. As it emits a burst of gorgeous light, the rocket disappears (for we never see the debris, the fragments of paper). The decay of a pion in flight is the same (except that there is no debris).

Do not let the debut of the pions overwhelm you. They are just a realistic replacement for something that today we can only imagine: a flashlight moving at $0.99c$.

8.4 The contrast

A contrast cries out to be made. Let me pick up a can of tennis balls, stand in front of you, and throw one horizontally. The speed of the ball relative to me is 30 miles per hour, say. Because I am not moving relative to you, the ball travels at 30 miles per hour relative to you, too. Part (*a*) of figure 8.3 depicts this.

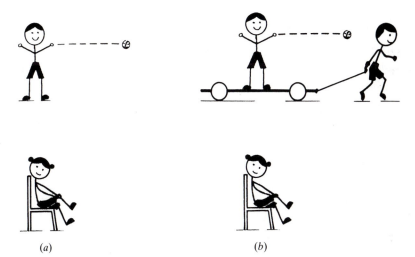

(*a*)   (*b*)

Figure 8.3 Thinking about the velocity of a tennis ball in two situations. (*a*) When throwing the ball, I am at rest with respect to you. (*b*) Now, when throwing the ball just as hard as before, I am moving with respect to you.

Now let someone pull me on a cart so that I move at 10 miles per hour relative to you. Part (*b*) of the figure shows this. Taking another ball from the can, I throw just as hard as before, and so the second ball goes 30 miles per hour relative to me and the cart. What about the speed of the second ball relative to you?

We might try the relationship

$$\left(\begin{array}{c} \text{Speed of ball} \\ \text{relative to you} \end{array} \right) \overset{?}{=} \overset{?}{} \left(\begin{array}{c} \text{speed of ball} \\ \text{relative to me} \end{array} \right) + \left(\begin{array}{c} \text{speed of me} \\ \text{relative to you} \end{array} \right)$$

$$? = ? \quad 30 \quad\quad + \quad 10.$$

The symbol ?=? means that we are asking whether the quantities on the left and right sides are actually equal. The proposed equation agrees with commonsense, and, indeed, an experimental test would find the speed relative to you to be very close to 40 miles per hour.

But what about light and the pion? The pion replaces me; the gamma rays replace the tennis balls; and the laboratory near Geneva replaces you in your seat. The analog of the tentative equation becomes

$$\begin{pmatrix} \text{Speed of light} \\ \text{relative to lab} \end{pmatrix} \quad ?=? \quad \begin{pmatrix} \text{speed of light} \\ \text{relative to now-defunct pion} \end{pmatrix} + \begin{pmatrix} \text{speed of pion} \\ \text{relative to lab} \end{pmatrix}$$

$$3 \times 10^8 \qquad\qquad ?=? \qquad 3 \times 10^8 \qquad\qquad\qquad + \begin{pmatrix} \text{greater than} \\ 0.99 \times 3 \times 10^8 \end{pmatrix}$$

The sum on the right-hand side gives a number close to 6×10^8 meters/second, but that is not at all what was found by experiment. For light, the commonsense way of combining velocities does not work.

Of course, we know that light is not like a tiny tennis ball. Newton's particle theory of light was based on such a picture, and the theory failed. To be sure, the photon is a particle of light, but it does not possess all the essential properties we ascribe to a tiny tennis ball. For example, we cannot ascribe to a photon a definite, continuous trajectory. Nonetheless, when we try to combine velocities, we are surprised to find that light and tennis balls present such a great contrast.

In the following chapters, almost everything that we find to be strange is traceable to the behavior of the speed of light. But we must be careful. Is "the behavior of the speed of light" a *cause* of the strangeness or an indication of something novel at a deeper level, specifically, at the level of the properties of space and time? This is a question to bear in mind.

Additional resources

Two reports describe in detail the experiment of this chapter: T. Alväger and others, *Physics Letters*, volume 12, pages 260–62 (1964), and T. Alväger and others, *Arkiv För Fysik*, volume 31, pages 145–57 (1966). The authors set the experiment in the context of direct tests of the principles of relativity theory.

The X-rays emitted by certain binary stellar systems provide additional excellent evidence that the speed of light is independent of the motion of the source. Kenneth Brecher analyzes the evidence in *Physical Review Letters*, volume 39, pages 1051–54 and 1236 (1977).

9 The principles of the Special Theory of Relativity

At last it came to me that time was suspect!

Albert Einstein,
conversation with R. S. Shankland, 4 February 1950

9.1 The big picture

Space and time are different from what you thought they were like. This simple statement is what the rest of the book is about. Right now, of course, the statement is enigmatic: how are space and time different? That will emerge by stages. The point here is to alert you to the big picture: when we follow Albert Einstein in developing the special theory of relativity, we are developing a theory of space and time. All of us have some commonsense notions about space and time, but – as we will discover – those ideas are not always valid.

This is sufficient prelude; let us go on.

9.2 The two principles

You sip from your coffee and then put the cup down. In the seat on your right, a woman pecks busily at the keyboard of a portable computer held in her lap. From what little you can see of the screen, you judge she is composing a sales report. On your left, music fills the ears of a 14-year-old boy. He leans back, his eyes closed and his cassette player languidly held in one hand. You pick up your coffee and have another sip.

Where are you? Flying at 500 miles per hour over Kansas on your way to Phoenix? Or still in the Chicago air terminal, delayed because snow has to be plowed off the runway? From what I have told you, you cannot distinguish between these alternatives – and therein lies the significance.

If you are in a uniformly-moving vehicle, its motion has no effect on the way things happen inside it. You lift your cup and swallow your coffee the same way in an airplane and in the terminal building. A computer and its TV-like display work just the same in both places. A cassette player reads information from a magentic tape and turns it into audible music – sweet or dissonant – the same way in a jet going 500 miles per hour and in the airport waiting room.

Various laws of physics govern the way computers and cassette players operate. The laws of electromagnetism govern the way a computer fetches information from its electronic memory. Those laws and the laws of motion govern the way an electron beam writes characters on the screen. The operation of a cassette player is subject to those laws, too. Because the computer and tape player operate the same way in the airplane and in the terminal, we infer that the laws of physics are the same in those two places.

To be sure, we should not take the foregoing to mean that we can never distinguish between flying and waiting in the terminal. If the plane runs into turbulent air, coffee will be spilt. (If you have ever tried to drink from a cup in a moving car, you know that a bump in the road or a sudden need for the driver to put on the brakes will put coffee in your lap.) When the airplane accelerates in its take-off, you feel the seatback pushing you, making you accelerate and thus ensuring that you keep up with the fuselage. This experience is very different from both cruising flight and waiting in the terminal.

In the examples of the preceding paragraph, we recognize instances of *non*-uniform motion. There is change in velocity; in technical language, there is *acceleration*.

We need to distinguish between *un*accelerated motion and accelerated motion. In unaccelerated motion, the velocity has a constant value. That constant value may even be zero; so unaccelerated motion includes the case of no motion at all. Another name for unaccelerated motion is *uniform motion*.

The tale that started this section provides us with a principle:

> if we are in an unaccelerated vehicle, its motion has no effect on the way things happen inside it.

We may state this principle in another, equivalent way. When you are flying, you take the rows of windows and seats, the cabin roof, and the carpeted floor as your frame of reference. For example, you note where the emergency exits are located relative to the window seat that you occupy. If you are still in the terminal, then other rows of seats, a ceiling, and a floor form your frame of reference. You note where the gate to the plane is relative to your current seat. Whenever we are awake, we have some frame of reference for describing the spatial location of objects, be they emergency exits or gates or coffee cups or computers or cassette players.

Our introductory tale pointed out to us that the laws governing the behavior of a computer, say, are the same in the cruising plane and in the terminal. So an equivalent statement of principle is this:

> the laws of physics are the same in all unaccelerated reference frames.

To describe in detail the behavior of a specific electron or a specific light beam, say, we need to use some reference frame. The principle says that, no matter which unaccelerated reference frame we adopt, we will use the same laws of physics to analyze the behavior. In short, the laws have the same structure no matter which unaccelerated reference frame we adopt.

The principles of the special theory of relativity

Now we are ready to state concisely the two principles on which Albert Einstein founded the special theory of relativity. Here they are.

Principle 1
Colloquial statement: if we are in an unaccelerated vehicle, its motion has
  no effect on the way things happen inside it.
Formal statement: the laws of physics are the same in all unaccelerated
  reference frames.

Principle 2
The motion of light is not affected by the motion of the source of light.

The first principle has two equivalent versions, one colloquial and the other formal. In any application of the principle, we may use whichever version is the handier.

The two principles – 1 and 2 – are the foundation on which we will build a theory of space and time. Some of our deductions will be bizarre and contrary to commonsense; so we should pause and ask, why should we adopt these principles?

For Principle 1, we have the tale with which this section started: things happen the same way in a uniformly-moving airplane as they do in the airport.

For Principle 2, we have evidence from the experiment described in chapter 8. The speed of the gamma rays emitted by a pion in swift flight is the same as the speed of visible light from a stationary lamp or laser. Thus Principle 2 is a generalization from excellent experimental evidence.

9.3 The logic of our development

At this point you may wonder, why does the topic we are studying carry the name "the special theory of relativity"? The name has three substantial words, and so the answer has three pieces.

The word "special" in the name arises because we employ only unaccelerated reference frames, not all reference frames that one can think of. In other words, we "specialize" to the way things appear when observed from uniformly-moving reference frames.

The specialization may leave you with questions; so let me add some comments here. Within the special theory of relativity, we may indeed describe and analyze *objects* that are accelerated, such as a baseball that is hit by a bat, but we restrict ourselves to doing so from *reference frames* that are uniformly moving. To admit for use all conceivable reference frames would be to undertake too large a task. That wider domain is the purview of Einstein's general theory of relativity.

The word "relativity" comes from a phrase coined by Henri Poincaré, an eminent French physicist and mathematician. In 1904, Poincaré was invited to address the International Congress of Arts and Science, held in St. Louis to commemorate the 100th anniversary of the Louisiana Purchase. Poincaré spoke of a "Principle of Relativity." To understand what Poincaré had in mind, let us look first at figure 9.1.

We see your airplane on its way from Chicago to Phoenix, another plane making the return flight, and the wheat fields of Kansas. A farmer, looking

Figure 9.1 An illustration of relative motion.

up, notes that you are flying southwest at 500 miles per hour relative to his wheat fields. The pilot of the return flight notes that the distance between the two planes is decreasing at about 1000 miles per hour. So far as that pilot is concerned, you are traveling at about 1000 miles per hour relative to his plane. We recognize these statements about your motion as meaningful. The essence is this: statements about uniform motion *relative to a specified reference frame* – wheat fields or another airplane – are meaningful.

A quantitative statement about uniform motion *without* specification of a reference frame is not meaningful. Why? Because our Principle 1 says we cannot discern uniform motion without recourse to some reference frame. Take first the colloquial form of that principle: if we are in an unaccelerated vehicle, its motion has no effect on the way things happen inside it. So by just doing experiments *inside* the vehicle, we have no way to assign a velocity to the vehicle. Only if we look out the window – and thereby use the wheat fields of Kansas as an outside reference frame – can we decide on a velocity (relative to that outside reference frame). The same conclusion follows if we use the formal version of Principle 1.

What Poincaré had in mind was this: when we deal with uniform motion, only motion *relative* to one or another specified reference frame is meaningful. This is the essence of Poincaré's Principle of Relativity. In the preceding paragraph, we saw that this essence follows from our Principle 1.

This paragraph is an aside, but you might like to read what Poincaré literally said. In 1904, Poincaré listed the major principles of physics; among them was

> the principle of relativity, according to which the laws of physical phenomena should be the same, whether for an observer fixed, or for an observer carried along in a uniform movement of translation; so that we have not and could not have any means of discerning whether or not we are carried along in such a motion.

You find the same content as in our Principle 1, and to that content Poincaré gave the name "the Principle of Relativity."

Let me put things in a nut shell. Both Poincaré and Einstein refer to our Principle 1 as "the Principle of Relativity." The "relativity" arises indirectly: Principle 1 implies that when we deal with uniform motion, only motion *relative* to one or another specified reference frame is meaningful.

The other noun in the name, the word "theory," appears because Principles 1 and 2 are generalizations from observation and experiment. As generalizations, they may be incorrect or too sweeping.

Indeed, the last point – about the fallibility of generalizations – suggests

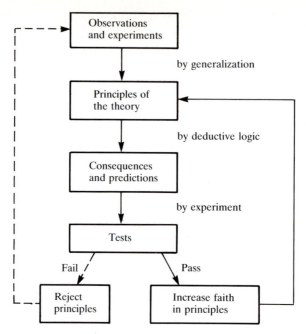

Figure 9.2 The logic of our development. There will be no need to follow the dashed lines.

that we be explicit about the way we develop the special theory of relativity. Figure 9.2 displays the logic. From observations and experiments, we go by generalization to the principles. Deductive logic – the sure logic – enables us to deduce consequences and make predictions, provided the principles are valid. Experiment enables us to test the predictions. If the predictions fail, we must go back to square one. If the predictions are confirmed, we are justified in increasing our faith in the principles, and we may cycle back to deduce further consequences.

9.4 The constancy of the speed of light

We make now our first deduction from Principles 1 and 2. The context is two reference frames in uniform relative motion. We may think of them as depicted in figure 9.3, literally two wooden frames, one labeled RF (for Reference Frame) and the other labeled RF′, the prime serving to distinguish it from the first frame. The arrow and letter v indicate that frame RF′ moves with speed v to the right relative to frame RF. A flashlight is at rest at the origin of frame RF. Its beam of light is sufficiently broad that

Figure 9.3 Reference frame RF′ moves with speed v relative to frame RF. Each frame has an observer in it (who is, of course, at rest with respect to that frame). A flashlight is at rest in frame RF.

experiments can be done on the beam in both frames. (The light is going through vacuum, not through air or glass or water or anything like that. To show evacuated tubes for the light to travel through would clutter the diagram.)

The question we ask is this: do the observers, Bob in frame RF and Alice in frame RF′, note different speeds for the light in the beam?

Right now we have a single source of light and two observers (in relative motion). What assistance can our principles give us in answering the question?

Principle 2 would tell us that Bob in frame RF would measure the same speed regardless of whether the flashlight is at rest in his frame or moving relative to it. But that is of no use to us because the flashlight is firmly at rest in his frame. The principles say something, but not what we need.

To be able to use the principles effectively, we must introduce an auxiliary light source: a flashlight at rest in Alice's frame that is identical in construction to the flashlight at rest in Bob's frame. Figure 9.4 shows the auxiliary flashlight in place. And we can reason as follows.

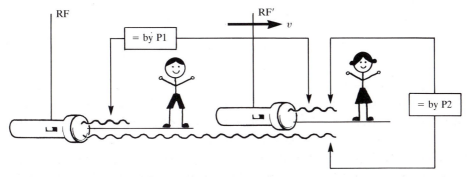

Figure 9.4 Now identical flashlights are at rest in each frame. The boxes indicate equalities in measured speeds and the reasons why.

By Principle 2, Alice must measure the same speed for the light from her flashlight and from Bob's flashlight, for the only difference is that his flashlight is moving (as perceived by her), but Principle 2 says that the motion of light is not affected by the motion of the source of light.

By Principle 1, Alice must measure the same speed for the light from her flashlight as Bob measures for the light from his flashlight, for the observers are doing identical experiments in their reference frames. (To be sure, Bob says Alice is moving (relative to him), and Alice says Bob is moving (relative to her), but such motion is without consequence here. For an observer in an unaccelerated reference frame, Principle 1 asserts that the frame's motion has no effect on the way things happen inside it. Therefore Alice and Bob must find identical speeds for the light from flashlights at rest in their respective reference frames.)

The next step is more easily intuited from the diagram than expressed in words, but here it is: the speed that Bob measures for the light from his flashlight and the speed Alice measures for the light from his flashlight are both equal to the speed she measures for the light from her flashlight. Therefore the first two speeds must be equal.

In short, although Alice moves with speed v relative to Bob, she notes the same speed for the light from the original source that he does.

The constancy

Our result for Bob and Alice generalizes readily. We have deduced that

> Observers in all unaccelerated reference frames measure the same speed for light (in vacuum) from any given source. That common speed is always c, where
>
> $c = 2.997925 \times 10^8$ meters/second
>
> $\approx 3 \times 10^8$ meters/second.

Succinctly stated, they *all* measure 3×10^8 meters/second *always* for light in vacuum. This remarkable property is called the *constancy of the speed of light*.

Comments

Some comments are in order.

First, note that both principles – Principle 1 and Principle 2 – are required to derive the conclusion, the constancy of the speed of light. The result is not inherent in either principle separately.

Second, our conclusion, remarkable though it is, refers only to the *speed* of light. The two observers do not find all aspects of the beam from Bob's flashlight to be the same. For Alice, the light from Bob's flashlight has a wavelength, a frequency, and an intensity that differ from what Bob notes. We need not delve into those differences; they are mentioned only lest you wonder whether everything is perceived to be the same. No.

Third, the constancy refers only to the speed of light in vacuum. The speed in glass or water is another issue, one that can be addressed only in chapter 13.

Fourth, the speed v with which Alice's frame moves relative to Bob's frame is specified to be less than c. Prudence suggests this stipulation. We have experience – through cars and airplanes – only of reference frames that move relative to each other at speeds much less than c. The speed of light, however, is emerging as a special speed. It may pose some kind of a limit, and chapter 13 addresses that issue. For now, we will be cautious and specify that the speed v may range from near zero to as large a fraction of c as one wishes. This restriction applies everywhere in chapters 9 through 12.

9.5 Synchronizing clocks within a single reference frame

Later in this chapter, and in other chapters, we will need to analyze events that occur at different locations within a single reference frame. For example, whenever we measure the velocity of a ball or an electron, we need the starting and finishing times at the two ends of some distance traveled. It is handy to have a clock right there where each event occurs. The clocks will be of identical construction and at rest in the single reference frame, but, even so, we need to be sure they are synchronized. For example, it would not do to have some clocks on standard time and others on daylight-saving time.

Here are two ways in which we might synchronize a set of clocks.

(1) Collect all the clocks in one spot, synchronize them, and then move them – slowly and carefully – to their distant locations.

(2) Place the clocks right away at their final locations. Use light signals to synchronize them.

Figure 9.5 shows how the second method would work. Clock A is the basic clock, and clock B is to be synchronized with it. From clock A, send a burst of light toward clock B and note the time (on clock A) when the light was sent: $t_{\text{sent out}}$. When the burst arrives at clock B, note the time that clock B reads and reflect the light burst back to clock A. When the burst returns to clock A, note (on clock A) the time of return: t_{return}. The round-trip travel time for the light is $t_{\text{return}} - t_{\text{sent out}}$, and half of that we can declare to

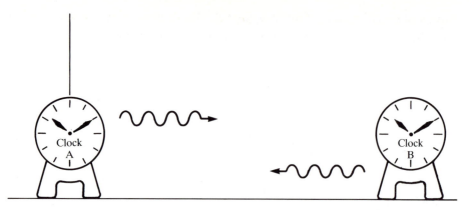

Figure 9.5 Synchronizing two clocks with a burst of light. The burst travels from clock A to clock B and is immediately reflected to clock A. The sketch shows the burst on both its outward journey and its return journey.

be the amount of time required to get from clock A to clock B. Thus we would like clock B to read

$$t_{\text{sent out}} + \tfrac{1}{2}(t_{\text{return}} - t_{\text{sent out}})$$

when the burst reaches it. If clock B did read that numerical value, then all is well already. If it did not, then we set clock B ahead or back by the requisite amount of time. (And then we do the steps all over again to make sure we got it right.)

We adopt the second method of synchronizing clocks. It is conceptually cleaner because we know so much about light, and it is also the choice made by Albert Einstein. Much later, in chapter 12, we will see that the first method would have led to the same results (in the limit of extremely slow transport of the clocks to their destinations).

You should not think of clock synchronization as some mysterious part of relativity theory. For convenience, we need several clocks. They have to be synchronized, and either of the two methods that seem obvious will do the job.

(Here is a parenthetical note, for the deeply critical reader only. Deferring synchronization to this stage in our development of relativity theory is logically acceptable, that is, synchronization is not a logical prerequisite for anything earlier in the text. In particular, one could develop experimental evidence for Principle 2 using non-synchronized clocks, one clock on standard time and the other on daylight-saving time, separated by 30 meters, say. Using those clocks to find out how long it takes light to go 30 meters would give a wildly incorrect value for the speed of light, but one could still establish that the speed had the same value, no matter whether the light source was at rest in the reference frame or moving.)

9.6 The relativity of simultaneity

You were born at some specific location and at some specific time. For your parents and for you, it was an event – in the colloquial sense of that word. Physicists use the word "event" in much the same sense: an *event* is anything that happens at some definite location at some definite time. Prototypical examples are your birth, the assassination of Abraham Lincoln, and your first Fourth of July firecracker. (In contrast, a forest fire that sweeps across 10 000 acres in five days does not constitute an "event" because the fire is spread out in space and time. The adjective "definite" means "distinct" or "limited" for anyone observing the happening. If the fire started when lightning struck a dead white pine, that inception would constitute an "event.") In this section, we begin to explore how events depend on the reference frame from which they are observed.

Figure 9.6 shows the two reference frames that we used earlier, while establishing the constancy of the speed of light. Two wooden markers, labeled (*a*) and (*b*), are permanently fixed in Alice's frame RF′. She places a firecracker mid-way between the two markers and ignites it. The firecracker explodes, sending light in opposite directions toward the two markers. Ultimately, the light will reach each marker, and those events – the reception of light by the markers – are the events that concern us. How do Alice and Bob perceive those events?

As observed by Alice in frame RF′

Light is emitted in both directions from a point *mid-way* between markers (*a*) and (*b*), *fixed* in Alice's frame of reference. She finds that light reaches

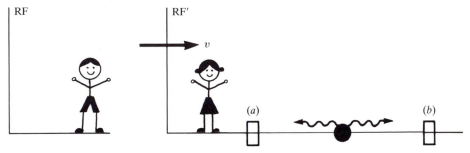

Figure 9.6 For Alice, the firecracker explodes mid-way between markers fixed in her frame of reference. Light reaches the two markers simultaneously – as perceived by Alice.

marker (*b*) at the same time that light reaches marker (*a*). For Alice, the two reception-of-light events are simultaneous. (Note that Alice may have assistants spread out in space, one at each marker. When light reaches a marker, the assistant stationed there notes the time on a clock present there and later reports the time to Alice. In working out the implications of Principles 1 and 2, cost is not a consideration.)

As observed by Bob in frame RF

Figure 9.7 illustrates what Bob observes. Light is emitted from a point *midway* between markers (*a*) and (*b*). Marker (*a*) moves *toward* the point where light was emitted; marker (*b*) moves *away*. The distance from source to (*a*) decreases; the distance to (*b*) increases. So the leftward traveling light (moving at speed *c*) reaches marker (*a*) before the other burst of light (also moving at speed *c*) reaches marker (*b*).

The event of reception at marker (*a*) occurs sooner than the event at marker (*b*) because the light does not have as far to go to reach marker (*a*) – as observed in frame RF. Therefore the two events are *not* simultaneous for Bob.

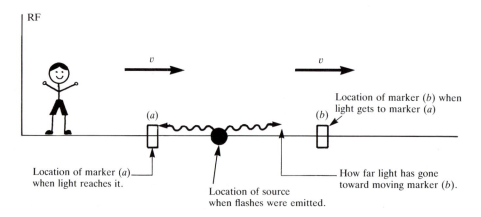

Figure 9.7 How Bob perceives the course of events. The arrows above the markers indicate their motion relative to frame RF. For Bob, the two reception-of-light events are *not* simultaneous.

Conclusions

From this exercise, we can extract a generalization:

> Spatially-separated events that are simultaneous in one frame are, in general, not simultaneous when viewed from another reference frame.

In turn, the generalization implies that there is no such thing as an absolute, global "now." If you say that a distant event is simultaneous with an event here and "now," I may legitimately disagree. And vice versa. For example, we could take reception-of-light at marker (*a*) to be here and "now." The reception-of-light at marker (*b*) is the distant event. If you perceive the events from Alice's frame and I from Bob's, then the events are simultaneous for you but not for me. That utterly undermines the commonsense belief that a universal "now" exists, independent of reference frame. The simultaneity of events is a relative notion, relative to the reference frame from which the events are observed. How peculiar!

9.7 Albert Einstein

This section provides a biographical sketch of Albert Einstein. Perhaps it is premature, in that we have only begun to study the special theory of relativity, and so you know little of its implications. If, however, you find yourself reading some of Einstein's non-technical essays and letters while you study the physics in the succeeding chapters, then this brief biography will have provided an historical context for that other reading. Indeed, browsing through the Einstein anthology *Ideas and Opinions* is a marvelous way to accompany the next five chapters.

Alienation

The Luitpold High School in Munich ran on rigid, authoritarian lines – it exemplified the worst of the German school system in the last decade of the nineteenth century – and young Einstein hated it. He had enrolled at age ten (a normal age for the school system), having previously attended a Catholic elementary school. Five years later, when his father's business failed and the family moved to Italy to start afresh, Einstein was left behind to finish up and get his high school diploma. Loneliness piled on top of academic alienation proved to be too much. A few months before he would turn sixteen, Einstein dropped out of school and left for Italy to rejoin his family. With him, he carried a letter from his mathematics teacher, vouching that Einstein's mathematics was already at university level – but he carried no diploma.

The next year was spent hiking, visiting museums, playing music, and reading – a time to discover and to learn freely. But the requirements of a career could not be deferred indefinitely. Planning to become an electrical engineer, Einstein took the entrance examination for Switzerland's equivalent of MIT, the Federal Institute of Technology in Zurich. He failed the exam. To be sure, he passed in mathematics and physics, but not in languages and biology.

Fortunately, examination was not the only way to gain admission. If Einstein got a Swiss high school diploma, admission would be automatic. So he went off to a progressive school in Aarau, lodged with the family of one of his teachers, and found a school atmosphere much more to his liking. Indeed, the experience transformed his career goals: henceforth he would be a secondary-school teacher of science and mathematics.

This period records, too, a change in Einstein's citizenship, the first and decidedly not the last. When he was born in Ulm, Germany, on 14 March 1879, he became, of course, a German citizen. The freedom permeating Switzerland stood in such painful contrast to Germany's authoritarianism that Einstein, age 16, renounced his German citizenship. Until he acquired Swiss citizenship five years later, he was a man without a country.

Admission to the Institute in Zurich brought another kind of alienation. As Einstein wrote in his "Autobiographical Notes,"

> I had excellent teachers (for example, [the mathematicians] Hurwitz and Minkowski), so that I really could have gotten a sound mathematical education. I worked most of the time, however, in the physical laboratory, fascinated by the direct contact with experience. The balance of the time I used in the main in order to study at home the works of [the physicists] Kirchhoff, Helmholtz, Hertz, etc.

That Einstein cut classes wholesale is proverbial. A few courses held his interest intensely, but often he found that the lectures were not on the portions of physics that fascinated him – and he stopped going to class. What is less often realized is that Einstein spent his time in the lab and in reading the primary literature, the research literature of physics. Einstein's route was not the easy way out. The research literature is written for other professional physicists, not for a student; steps are omitted, and the material is difficult. But there Einstein could find the developments, especially Maxwell's theory of electromagnetism, that fascinated him.

Overall, there was a serious mis-match of aims. Einstein was eager for research on the most distant frontiers of physics, and in some ways he was ready, for he had ideas. The Institute, however, wanted to establish a solid background in physics, especially in someone who was enrolled as a prospective high school teacher.

By his own methods, Einstein grew in knowledge and sophistication, but there was a price to pay. Again, as he related in his "Autobiographical Notes,"

> In this field [physics], however, I soon learned to scent out that which was able to lead to fundamentals and to turn aside from everything else, from the multitude of things which clutter up the mind and divert it from the essential. The hitch in this was, of course, that one had to cram all this stuff into one's head for the examinations, whether one liked it or not. This coercion had such a deterring effect [upon me] that, after I had passed the final examination, I found the consideration of any scientific problems distasteful to me for an entire year. In justice, I must add that in Switzerland we had to suffer far less under such coercion, which smothers every truly scientific impulse, than is the case in many another locality. There were altogether only two examinations; aside from these, one could just about do as one pleased. This was especially the case if one had a friend, as did I, who attended the lectures regularly and who worked over their content conscientiously. This gave one freedom in the choice of pursuits until a few months before the examination, a freedom which I enjoyed to a great extent, and I gladly took into the bargain the bad conscience connected with it as by far the lesser evil. It is, in fact, nothing short of a miracle that the modern methods of instruction have not yet entirely strangled the holy curiosity of inquiry; for this delicate little plant, aside from stimulation, stands mainly in need of freedom; without this it necessarily goes to wreck and ruin. It is a very grave mistake to think that the enjoyment of seeing and searching can be promoted by means of coercion and a sense of duty.

Thus, with the aid of Marcel Grossmann's lecture notes, Einstein graduated from the Institute in 1900. He was then 21 years old, was certified to teach mathematics and physics at the high school level – and was alienated once again, this time from physics itself.

Einstein proceeded to support himself with an assortment of jobs: two temporary teaching positions, a little tutoring, and the like. He looked for a regular job but could not get one. The reason, in part, was his cavalier attitude while at the Institute: he had alienated – it goes in the other direction now – those who could write letters of recommendation for him. Grossmann's father came to the rescue, getting Einstein an interview with the director of the Swiss Patent Office in Bern. The interview left the director with mixed feelings: the young man lacked the relevant technical qualifications for an examiner of patent applications, but there was something peculiarly intelligent about him. The director offered Einstein a provisional job – and even that only when the next opening should happen to arise. Einstein actually started work at the Patent Office in June of 1902.

The Patent Office years

Einstein's new job entailed reading patent applications, figuring out what the essentials of the purported invention were, and – if necessary – rewriting the application to make that clear. The task was not onerous, and Einstein found time to work on his own research at the office. One drawer of his desk was dedicated to that. Before coming to the Patent Office, he had published two papers: one on capillary phenomena – perhaps most familiar to you as the way soda rises slightly in a straw even before you sip – and another on the chemical thermodynamics of metals and molecular forces. The next two years saw more papers on thermodynamics, about one per year, papers focussed on securing the foundations of the subject.

Physics was not, by any means, all of Einstein's life. In 1903 he married Mileva Marić, a classmate of his at the Institute. Their marriage was blessed with two sons, but it was not a marriage destined to last a lifetime.

For a sense of historical perspective, let us note that Einstein received his college degree in the year – 1900 – that Max Planck discovered what came to be called Planck's constant, a discovery that – slowly but certainly – undermined the classical physics of Newton and Maxwell. Moreover, although Maxwell and Hertz had established an electromagnetic wave theory of light, it was not clear how that theory should be expressed in various frames of reference. The issue was taxing the best brains in physics – among both experimenters and theorists – as the nineteenth century turned into the twentieth. Einstein was growing into scientific proficiency at a most auspicious time. And in 1905 he published four remarkable papers, each of them world-class.

The four papers were preceded in that year by a paper entitled "A new determination of the size of molecules." In those days, the very existence of atoms and molecules was doubted by certain quite renowned physicists and chemists. Any new method to determine sizes by experiment – and to show consistency with other methods – lent credibility to the atomic hypothesis. Moreover, Einstein was able to submit this work to the University of Zurich and receive a doctorate in return.

The first world-class paper of 1905 was Einstein's paper on a particle theory of light, what came to be known as the photon theory of light. We discussed this paper in section 6.4 and noted there that it provided the grounds – nominal, at least – for Einstein's Nobel Prize.

The next paper was on molecular motion (rather than molecular size) and its connection with Brownian motion, the erratic zig-zag motion of pollen grains in water or of dust specks in air. Einstein showed how the erratic motion of the dust speck is related mathematically to molecular impacts by air molecules. His derivation provided another way to determine explicitly the size of molecules, a project brilliantly executed in the succeeding years by the French physicist Jean Perrin.

Then came the most memorable paper – the special theory of relativity – but under an unprepossessing title: "On the electrodynamics of moving bodies." The title reflects the preoccupation of the day: how to describe electromagnetic phenomena – electricity, magnetism, and light – from various frames of reference. The body might be at rest in one such frame but moving when observed from another; from this circumstance arises the concern with "moving bodies."

The fourth of the world-class papers is Einstein's first derivation of his best-known equation: $E=mc^2$. Chapter 11 is devoted entirely to this equation, its implications, and some of its history; we can leave the equation for now and dwell a bit on the relativity paper.

Einstein's 1905 relativity paper is curiously analogous to Lincoln's Gettysburg Address. The funeral orations of Lincoln's day were lengthy affairs. Edward Everett, formerly both a US senator and a president of Harvard, spoke at Gettysburg for two hours. When he had finished, Lincoln rose, spoke the ten sentences of his address – it took less than three minutes – and sat down. The audience did not know what to make of it; yet today we recognize the address as a classic tribute, moving, deep, and immortal.

So it was with Einstein's paper. The community of physicists was accustomed to lengthy, complex calculations based on detailed dynamics, an intricate mathematical combination of Newton's laws of motion and Maxwell's theory of electromagnetism. Einstein said brashly, wait, let's just take seriously two simple principles and see where they lead us, even if that means giving up our old notions of time and simultaneity. And then, with substantially simple mathematics, he worked out a host of consequences. Because Einstein approached the issue from so radically different a point of view, the community of physicists paid scant attention to his paper. It wasn't in the mainstream; it wasn't in the usual mode of attack on the problem. Only Max Planck – not coincidentally – saw merit in the paper of this unknown young man, recommended it to his colleagues, and proceeded to follow up on it in his own research.

Figure 9.8 shows Einstein as we would do well to think of him when he wrote the papers of 1905.

The blindness of the physics community is well illustrated by the follow-

Figure 9.8 Albert Einstein in Zurich, perhaps five years before 1905. The halo of white hair and the gentle smile came decades later. Rather, the young man who invented the special theory of relativity was brash and nonchalant, as befitted the task. (Courtesy of Lotte Jacobi Archive, University of New Hampshire.)

ing aside. In the Swiss and German academic systems, the route to a professorship started with the post of *Privatdozent*. The meaning is literally "private instructor" at the university, a teacher who gives lectures in return for the fees that students pay directly to him or her. To qualify for the post, a person must submit a significant piece of research and give a public lecture on the subject. (All this occurs after a doctorate has been earned and separately from it.) In late 1907, two years after the relativity paper had appeared, Einstein submitted that paper to the University of Bern as his research for the post of Privatdozent. The paper was rejected. "Incomprehensible" – no, not the university's action, but rather the verdict by the faculty on Einstein's paper.

Recognition, nonetheless, was beginning to grow. In the same year – 1907 – the mathematician Hermann Minkowski reformulated Einstein's relativity paper into a theory of space–time, a four-dimensional space. The content did not change, but it became more apparent – mathematically, anyway – and also more noticed by the communtiy of physicists. In 1908, the University of Zurich sought Einstein for its faculty. First, however, he must become a Privatdozent. Reluctantly, Einstein applied again to the University of Bern, was accepted as a Privatdozent, and commenced giving lectures (while still working at the Patent Office). In due course, the faculty position materialized – it was the equivalent of an associate professorship in America – and in May of 1909, Einstein became Herr Professor Doctor Einstein at the University of Zurich. He was thirty years old, and his years of obscurity were over. In another ten years, his years of privacy would be over, too, but that comes later. First we must follow Einstein around Europe and the world.

Cosmopolite

In 1909 Einstein left Bern and its Patent Office and moved to Zurich, its university, and his first professorial appointment. Soon he was sought by the German University in Prague, Czechoslovakia, then a part of the Austro-Hungarian empire. The post – a full professorship – required that the applicant have a recognized religious affiliation. This posed a problem for Einstein, though not an insuperable one. His parents, Jewish by descent, had not been devoutly religious. Young Albert had attended a local Catholic public elementary school in Munich. State law required religious instruction, and Albert received Catholic instruction in school and Jewish instruction at home; the latter was provided by a family relative (although Albert could have received it at another, nearby public school). The boy became deeply religious in a traditional sense – until the age of twelve, when his reading in popular science books convinced him that much of the Bible could not be true. When first approached about the post in Prague, he had declared himself "unaffiliated" religiously. That, however, was not good enough, and so he changed the designation to "Jewish." Einstein could do this in good conscience, for he felt himself ethnically and culturally Jewish, while simultaneously not being religiously Jewish. As he described himself years later,

> It seems to me that the idea of a personal God is an anthropological concept which I cannot take seriously. I feel also not able to imagine some will or goal outside the human sphere. My views are near those of Spinoza: admiration for the beauty of and belief in the logical simplicity of the order and harmony which we can grasp humbly and only imperfectly. I believe that we have to content ourselves with our imperfect knowledge and understanding and treat values and moral obligations as a purely human problem – the most important of all human problems.

While in Prague, Einstein was invited to the first international research conference sponsored by Ernst Solvay, a Belgian industrial chemist who had amassed a fortune by making sodium bicarbonate efficiently. Here was recognition indeed, for only twenty-one individuals attended the conference, among them Madame Curie and Henri Poincaré. Both would write letters of recommendation for Einstein when he went on to his next, more prestigious job: professor at his alma mater, the Swiss Federal Institute of Technology, a post he accepted in 1912.

Einstein's appointment at his alma mater was for a 10-year period, but the span was really irrelevant. In the summer of 1913, Max Planck and Walther Nernst, two of the most senior physicists in Europe, traveled from Berlin to Zurich to talk to Einstein and to convince him to accept a post in Berlin that would be established specifically for him. He would be elected to the Royal Prussian Academy of Science, be given the directorship of a brand-new physics institute, and be free to divide his time between teaching and research in any way he wished. The world of physics could not offer a position more prestigious or with greater opportunity for self-fulfillment. Einstein hesitated – he distrusted German militarism, and World War I was on the horizon – and then accepted the offer. In April of 1914, he and his family left Switzerland for Berlin.

The war broke out in August of that year. Einstein's wife Mileva and the two children went back to neutral Switzerland, but the move also signaled the marriage's end, though formal divorce would not come until five years later. Einstein took no part in World War I; we can indeed pause in this chronicle to catch up on Einstein's science since that wonder year of 1905.

The special theory of relativity is "special" because it is restricted to describing the way the world appears when viewed from reference frames in uniform motion. Soon after developing the special theory, Einstein realized that he could incorporate gravity in a natural and correct way only if he generalized his framework to include all conceivable frames of reference. For example, this means not only an airplane in smooth flight but also one tossed in the turbulence of a thunderhead. Or not only a car cruising along a straight stretch of Nevada salt flat but also a car accelerating around a curve at the Indianapolis 500. In the decade from 1905 to 1915, Einstein literally generalized his previous work, developing the general theory of relativity, a theory of both gravity and the structure of space and time. In finding the right mathematics to express the theory and in working with that mathematics, Einstein had the help of his friend Marcel Grossmann. First the borrowed lecture notes, then the Patent Office job, and now help with the esoteric mathematics – Grossmann was a valued friend indeed. The pair of them published two joint papers as Einstein groped his way toward the extended theory.

Now we examine two aspects of the general theory. We are accustomed to the idea that the sun's gravitational attraction pulls on the earth, thereby bending what would otherwise be a straight-line trajectory into a nearly circular trajectory. More precisely, the solar attraction bends the earth's trajectory into an elliptical form: the earth's orbit is an ellipse. This shape follows from Newton's laws of motion and the gravitational force law of his day. A similar elliptical path is predicted for each of the planets: Mercury, Venus, Mars, . . . and out to Pluto. Well, not quite. The elliptical path is the Newtonian prediction provided one neglects the gravitational attraction exerted on each planet by all the others. When this small, extra effect is included, the predicted shape of Mercury's orbit is still close to elliptical, but the trajectory does not quite repeat itself in radial location each time Mercury goes completely around the sun. The point of closest approach to the sun – the *perihelion* point – shifts a bit in angular location. Thus Newtonian theory predicts a shift of the perihelion, and observation confirms the existence of such a shift – but the numbers do not quite agree. This discrepancy had been an embarrassment to astronomy for half a century. Einstein's new theory of gravity gave a perihelion shift different from the Newtonian value – and in striking agreement with the astronomical observation. It was a subtle but brilliant achievement.

More dramatic in its development was Einstein's prediction that the path of starlight would be bent by the gravitational influence of the sun. The

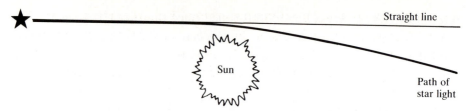

Figure 9.9 The gravitational bending of starlight by the sun. For clarity's sake, the curvature of the light path shown here is much exaggerated. In reality, the path of light remains very close to a straight line.

suggestion, illustrated in figure 9.9, was actually not a new idea. Calculations had been made already around 1784 by the English physicist Henry Cavendish, who never published his computation, and again in 1801 by the German astronomer Johann Georg von Soldner, who did publish his work. Their calculations were based on the idea of light as a particle in the sense of Newton's theory of light: a small body of ordinary matter. Einstein's calculation flowed from a wave theory of light – yes, despite his espousal of the photon idea – and it predicted a deflection twice as large as the previous, "Newtonian" computations. Einstein made his definitive prediction in the middle of World War I, when Germany and England were pitted in bloody battle. Yet the British astronomers, most notably Arthur Eddington and Frank Dyson, saw Einstein's general relativity and this prediction as something that could raise people above chauvinistic feelings. Eddington and Dyson set out to test the prediction during the solar eclipse of 1919, when the stars near the sun (in angular location) would be visible. Coming in the first year after the end of the war, the eclipse expedition vindicated Einstein's prediction – and catapulted him to public fame, fame such as no scientist before him or since him has attained. Perhaps Einstein and his successful, abstruse theory of gravity, space, and time were seen as a symbol of what is lofty about humans, so different from the horror they had wrought in the trenches of Europe.

The year 1919 brought other changes in Einstein's life. Early in that year, he and Mileva were formally divorced, and in June he married his cousin Elsa, a widow with two daughters. He and Elsa had known each other in their youth; she had nursed him through a serious illness toward the end of the war; and she would look after him until her death in 1936.

Germany's defeat in World War I cast the country into intense political chaos. Stunned and confused, some Germans looked for scapegoats – and found them among the Jews and the political left. A rising anti-semitism drove Einstein to support Zionism, while still dissociating himself from Judaism as a religion. He felt a oneness with every persecuted people and especially with the Jews, his cultural and ethnic brethren. His first trip to the United States, taken with Chaim Weizmann in 1921, was a trip to raise funds for the Hebrew University in Jerusalem.

The eighteen years from 1915 to 1933 were less productive than the remarkable decade that preceded them. Of course, only for a person like Einstein would one characterize the period as "less productive." He developed cosmological implications of general relativity, specifically, the idea that the universe, though without spatial boundaries, may nonetheless be of finite volume (much as the surface of a sphere, perceived strictly as a two-dimensional space, has no boundaries but does possess a finite area). Einstein worked further on the quantum theory of light but separated himself increasingly from the quantum theory of the atom as it was being developed in Göttingen, Copenhagen, Zurich, Cambridge, and even Berlin. Rather, he focussed his attention on further generalization of his general theory of relativity, seeking to extend its purview from gravity alone to gravity plus electromagnetism and whatever else might be fundamental in physics. Thus he sought what came to be called a "unified field theory," where the word "field" is understood in the sense in which we used it in chapter 5 and its idea of the electric field.

Indeed, the idea of a field was a guiding theme in Einstein's conception of a proper physics. In the study at his home, Einstein kept two portraits: Michael Faraday and James Clerk Maxwell. Faraday introduced the field into electromagnetism, and Maxwell developed the idea mathematically, with stunning results, as we learned in our discussion of electromagnetic waves. There is a curious coincidence here, too. Just as Newton was born the year Galileo died, so was Einstein born the year Maxwell died. Physics records a remarkable continuity.

When Adolf Hitler came to power in the winter of 1933, Einstein was in the United States, a visiting professor at the California Institute of Technology. (Its president, Robert Millikan, had confirmed Einstein's predictions about the photoelectric effect two decades earlier.) Einstein resigned from the Prussian Academy and from his job in Berlin, and he announced that he would not return to Germany. After brief sojourns in Belgium and England, Einstein moved in October 1933 to Princeton and its new Institute for Advanced Study, never to leave.

The Princeton years

Einstein's years in Princeton bring us the poignant pictures of the saintly, white-haired elder statesman. The era started, however, when Einstein was only 54 years old and in the prime of middle age. His efforts were devoted to two causes. First, he continued to search for a unified field theory. At the Princeton Institute, he had a succession of able mathematical assistants – post-doctoral fellows of a sort – but the search for a unified theory had no guiding physical principles such as the search for general relativity had had. One might as well come out and say it right away: although Einstein searched literally to his death bed in the Princeton hospital, he never found a successful generalization of his theory of 1915. He worked far from the mainstream of physics and criticized that stream incisively in 1935. He could not bring himself to believe in an inherent randomness for quantum phenomena, such as the irregular appearance of photons that we discussed in chapter 7. "*Gott würfelt nicht*" was his attitude: God does not play dice.

The second cause was the cause of world peace, broadly construed. Already in July of 1933, Einstein temporarily suspended his pacifism, saying that Hitler's Germany required that other European countries prepare to defend themselves militarily. He helped countless refugees from fascist Europe find a welcome and a job in America. In October of 1940, a year after World War II had broken out in Poland, Einstein became a citizen of the United States, a gesture of support for the democracies and the country that was his host.

Einstein's role in the race to build the atomic bomb is much less than most people realize, but the history of the bomb comes later, in chapter 11. During the war, Einstein did regularly consult for the Navy on conventional high explosives and ballistics.

In the years after the war, the proliferation of nuclear weapons and the drive to build the hydrogen bomb troubled Einstein immensely. He spoke out against what he saw as humanity's ultimate folly: a world living under the threat of nuclear annihilation.

In the closing years of his life, Einstein was offered the presidency of Israel. Though honored and deeply moved, he declined the offer. Only a few years were left for the frail elder statesman of physics: he died peacefully on 18 April 1955, remembered well by a world that had already revered him for decades.

Additional resources

Banesh Hoffmann provides a fine historical account of relativity theory in his little paperback *Relativity and Its Roots* (W. H. Freeman, New York, 1983). As a one-time assistant to Einstein, Hoffmann knows whereof he writes. The little book showed me that much of special relativity theory can be presented with diagrams, rather than with algebra, and for that insight I am forever grateful to Professor Hoffmann.

Einstein's "Autobiographical Notes" appear in the volume *Albert Einstein: Philosopher-Scientist*, edited by Paul Arthur Schilpp (Tudor Publishing, New York, 1949). The Einstein anthology *Ideas and Opinions* is available in paperback from Crown Publishers, New York (1985). Banesh Hoffmann and Helen Dukas, Einstein's secretary, provide a charming biography in *Albert Einstein, Creator and Rebel* (Viking Press, New York, 1972). The anthology *Einstein, A Centenary Volume*, edited by A. P. French (Harvard University Press, Cambridge, MA, 1979) is a treasure trove: reminiscences of Einstein by his associates, some of Einstein's own writings, and essays that set his work – both scientific and political – in context and assess its influence on the world.

Questions

1. Figure 9.10 shows a distant star and two rocket ships, my ship at rest with respect to the star and your ship moving toward it at a speed of $0.3c$. Do you and I measure the same speed for the light from the star?

The figure also shows two identical auxiliary light sources, one at rest in each rocket. Use them and our principles to answer the question above by reasoning much as we did in section 9.4.

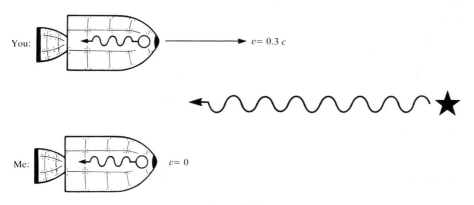

Figure 9.10

2. In section 9.6, we found that, for Bob, reception-of-light at marker (a) occurred before reception-of-light at marker (b). From Alice's point of view, Bob was moving away from her toward the left.

Suppose you are in a third reference frame FR'', moving with speed v away from Alice toward the right, as she perceives your motion.

(1) How do you perceive Alice's motion? (Sketch your frame and hers, too.)

(2) For you, which event occurs earlier, reception-of-light at (*a*) or at (*b*)?

(3) Can you adduce reasons to conclude that "if two spatially-separated events, at locations (*a*) and (*b*), are simultaneous in one frame, we can find frames in which the event at (*a*) precedes the event at (*b*) and other frames in which the event at (*b*) precedes the event at (*a*)."

(4) If the statement in part (3) can be supported, does it constitute a violation of causality?

3. Tacking in still air. Professor Wendell Furry was one of my undergraduate advisors. While I was in his office one day, a colleague came in and declared, "Wendell, I have a problem for you." Here is the conundrum.

Each of two individuals claims that his sailboat goes faster than the other's. To settle the dispute, they agree to race from one bridge on the Charles River downstream to the next.

On the appointed day they meet – but there is no wind. The air is perfectly still. Undismayed, they get into their boats and start drifting downstream.

Then one fellow notices a wind in his face. (He is, after all, drifting through the still air.) The wind is dead against him, but he says to himself, "I can tack upwind." And so he proceeds to hoist his sail.

Questions: Does the race have a winner? If so, who is it? And why?

(For those of you who are not sailors, I append the following definition: to *tack* is to maneuver a sailboat upwind, that is, toward the wind but at close to a 45° angle to it, as sketched in figure 9.11. To sail upwind, a sailor adroitly combines sail, keel, and rudder; it is a three-element effect.)

4. Experimental research on the shape of airplane wings is often done in a "wind tunnel." A model of the wing is held fixed on a pedestal, and a wind is blown past the wing. Any abnormal bending or fluttering is keenly observed.

In the real world, however, wings move, rather than being at rest. And they move through air that is largely quiescent.

What possible relevance can wind tunnel measurements have for "real world" engineering? And why?

5. Review the presentation in section 9.4, The constancy of the speed of light, and then respond to the following questions.

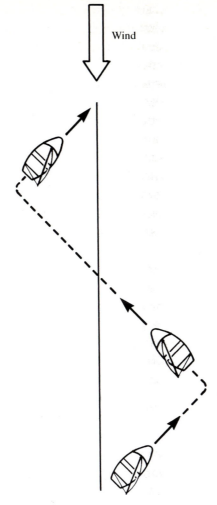

Wind

Figure 9.11 A view from above of a sailboat tacking.

Do you find the logic convincing? If not, to what do you object? If you accept the conclusion (provisionally, subject to experimental test), what changes in your conception of space and time may be necessary (to accommodate the constancy of the speed of light)?

6. Let's suppose you can juggle three tennis balls while standing in your bedroom. Would juggling them be any different if you were standing in the aisle on a train traveling 70 miles per hour between Philadelphia and Washington? Explain your response.

7. Bob wants to be sure to wake up at 8 a.m., and so he sets two alarm clocks, each placed on the table beside his bed. The next morning, as the hand on each clock reaches 8, an obnoxious buzzing and a loud ring propel Bob out of bed. For Bob, the clocks rang simultaneously.

(*a*) If Alice were traveling past Bob at nine-tenths of the speed of light, would she perceive the two alarm clocks to ring simultaneously? (In thinking about this question, you may regard "Bob's table" as a single spatial location, so that the two alarm clocks are at the same place.)

(*b*) The situation is different with two events that occur at different spatial locations (at least for some observers). Do you have any reason, based on your own experience, to expect that spatially separated events will be perceived as simultaneous by all observers if they are simultaneous for one observer?

8. Are you comfortable with abandoning the idea of an absolute, global "now"? If your response is "no," how would you try to defend a belief in its existence?

9. An international crisis loomed. In Washington, the Speaker of the House pounded a gavel to call Congress into session. Simultaneously, in New York, the Peruvian representative to the United Nations rapped a gavel to call the Security Council to order. What meaning ought the word "simultaneously" have here? Would all observers agree that the two events were indeed simultaneous?

10. "Sea sickness" and "car sickness" are sometimes lumped together under the rubric "motion sickness." Is this phrase a misnomer? Does motion cause the sickness? Can it? Should the affliction be called "change-of-motion sickness"?

11. In this chapter, we derived the constancy of the speed of light: a given burst of light has the same speed in all uniformly moving frames of reference. Can one derive, in some analogous fashion, the proposition that a baseball has the same speed in all uniformly moving frames? If yes, how? If no, what item in the analogy fails?

10 Time dilation and length contraction

Einstein concentrated on changing the traditional notions of absolute space and absolute time, replacing them by one absolute: the speed of light.

Anthony J. Adams, Wesleyan '91,
essay in Physics 104, fall 1988.

10.1 Time dilation

The relativity of simultaneity tells us that observers in different frames of reference may disagree on the time interval between two events. One observer may find the time interval to be zero and hence the events to be simultaneous; another may not.

Let us try to make the comparison of time intervals more quantitative. We suppose that Alice is baking brownies. She puts the tray in the oven and sets the timer for 30 minutes. This is the first event. When the timer rings, Alice takes out the brownies; that is the second event. For Alice, the time interval between the two events is 30 minutes, and the events occur at the same location: the oven, stationary in her frame.

As in chapter 9, Alice moves with speed v relative to Bob, and so do the oven and brownies. What time interval between the two events does Bob perceive?

To answer the question, we need somehow to introduce light, for all that we are sure of is that *both* Alice *and* Bob always measure the speed of light to be c, 3×10^8 meters/second. So let us imagine a mirror on the ceiling over the oven, as sketched in figure 10.1. When Alice puts in the brownies, she also sends a burst of light up toward the mirror. We can *imagine* the ceiling so high that light takes 30 minutes to reach the mirror, reflect, and return to the oven. Thus the events labeled 1 and 2 in the sketch coincide with the events of putting in the brownies and taking them out; we can concentrate on the emission and reception of the light burst.

For the time interval between events 1 and 2 as perceived by Alice, we use the symbol Δt_{Alice}. The capital Greek letter delta, Δ, is read as "change in," and so Δt_{Alice} is read as "change in Alice's time," meaning "the

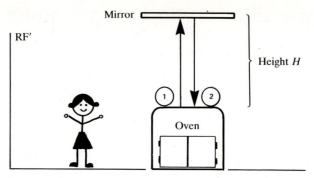

Figure 10.1 Alice sends a burst of light from oven to mirror and back to oven.

Event 1: burst of light leaves oven.

Event 2: burst of light returns to oven.

elapsed time as measured by Alice." We can express Δt_{Alice} in terms of the height H from oven to mirror. For Alice, the light makes a round-trip of length $2H$ at speed c. The product of speed and elapsed time gives the distance traveled, and so

$$c\,\Delta t_{\text{Alice}} = 2H.$$

Then division on both sides by c yields

$$\Delta t_{\text{Alice}} = \frac{2H}{c}.$$

Figure 10.2 shows how Bob perceives the two events. Because Alice,

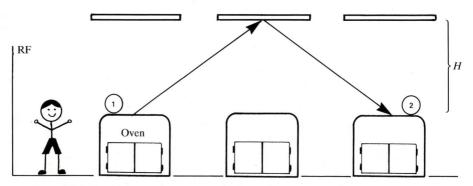

Figure 10.2 How Bob perceives the course of events. The oven and mirror move with speed v relative to Bob's frame. They are shown three times: when the light burst is sent out, when it reflects, and when it returns. (For future reference, note that – for Bob – events 1 and 2 occur at different spatial locations. This is a key distinction between what Alice and Bob perceive.)

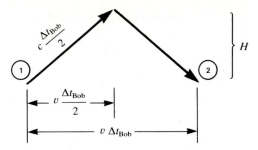

Figure 10.3 The geometry of the light paths as Bob perceives them.

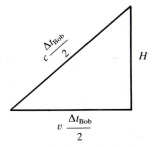

Figure 10.4 The crucial right triangle.

oven, mirror, and brownies are moving relative to Bob's frame, the light makes two diagonal paths, first diagonally up to the mirror and then diagonally back down to the oven.

Because light has farther to go (as Bob sees things) and because the speed of light is the same for Bob and Alice, Bob notes a longer time interval than Alice does. Denoting by Δt_{Bob} the time interval Bob perceives, we have the inequality $\Delta t_{\text{Bob}} > \Delta t_{\text{Alice}}$. This is another surprise.

With the aid of the Pythagorean theorem, we can be fully quantitative, as follows.

Figure 10.3 lays out the geometry in more detail. In the time interval Δt_{Bob}, the oven and mirror move a distance $v\Delta t_{\text{Bob}}$. At the instant when light is reflected from the mirror, half the full time interval has elapsed, and so the mirror (and oven) have traveled a distance $v(\Delta t_{\text{Bob}}/2)$. The light burst, traveling at speed c, has traversed a diagonal path of length $c(\Delta t_{\text{Bob}}/2)$.

Figure 10.4 shows the left half of figure 10.3 and thus provides a right triangle; we can use the Pythagorean theorem to extract Δt_{Bob} in terms of H, v, and c. Thus, to start with, the Pythagorean theorem gives us

$$\left(c\,\frac{\Delta t_{\text{Bob}}}{2} \right)^2 = \left(v\,\frac{\Delta t_{\text{Bob}}}{2} \right)^2 + H^2.$$

Subtract $(v\Delta t_{Bob}/2)^2$ from both sides and then factor on the left:

$$(c^2 - v^2) \left(\frac{\Delta t_{Bob}}{2} \right)^2 = H^2.$$

Multiply both sides by $2^2/c^2$:

$$\left(1 - \frac{v^2}{c^2} \right) (\Delta t_{Bob})^2 = \frac{2^2 H^2}{c^2}.$$

Take a square root:

$$\sqrt{1 - (v^2/c^2)} \; \Delta t_{Bob} = \frac{2H}{c}.$$

Now we have Bob's time interval expressed in terms of H, v, and c. For a clear comparison with Alice's time interval, we go on two steps. Because Alice's time interval is given by $\Delta t_{Alice} = 2H/c$, we substitute her time interval on the right-hand side:

$$\sqrt{1 - (v^2/c^2)} \; \Delta t_{Bob} = \Delta t_{Alice}.$$

Divide both sides by the square root:

$$\Delta t_{Bob} = \frac{1}{\sqrt{1 - (v^2/c^2)}} \; \Delta t_{Alice}.$$

Because the divisor is less than 1, Bob does indeed note a longer elapsed time than Alice does.

What intrinsically distinguishes the situations as perceived in the two reference frames? In Alice's frame, the two events occurred at the *same* place; in Bob's frame, where the stove moves, they did not. So we may generalize the result just derived to read as follows:

$$\Delta t_{\text{any other frame with relative speed } v} = \frac{1}{\sqrt{1 - (v^2/c^2)}} \; \Delta t_{\text{frame where events occur at } the \; same \; place}. \tag{1}$$

This is the relationship to bear in mind. Informative subscripts distinguish the two time intervals (rather than "Bob" and "Alice," whose significance one can easily forget). As for remembering why the relationship arises, the best thing is to remember the figures: Bob's diagonally upward and downward light paths are longer than Alice's vertically up and down paths, and so Bob perceives a longer time interval than Alice does.

Comments

We can state the qualitative significance of equation (1) as follows.

> The context is a pair of events. An observer for whom the events happen at the same place measures the least elapsed time.

(Parenthetical note: for an arbitrarily-given pair of events, we may not always be able to find such an observer or to imagine one legitimately, but whenever we can, the conclusion holds.) Other observers, for whom the events occur at two different locations, measure a longer time interval. This phenomenon carries the name *time dilation*.

The property displayed in equation (1) is an intrinsic property of space and time and relative motion. It is not an artifact arising because Alice's clock or Bob's clock does not keep time correctly. We can provide Alice and Bob with clocks of identical structure and the finest quality. The clocks may be hand-wound pocket watches in gold cases or digital watches that are waterproof to 50 meters. According to our Principle 1, the motion of the clocks (Alice's clock relative to Bob's and vice versa) has no effect on the way the clocks function and hence keep time. Alice's measurement is just as valid as Bob's and vice versa. No one's clock is "running slow" in the sense that the clock is malfunctioning and should be returned to the store. Rather, we are forced to conclude – once again – that the time interval between two events is a relative quantity, relative to the frame of reference from which the events are observed.

In short, the fault is not with the clocks. It is with our intuition. We expect all observers to measure the same time interval between a pair of events, but space and time are not that way.

Thus far we have analyzed in detail two pairs of events, one pair in section 9.6 and then another pair in the present section. Table 10.1 displays the essential elements and the conclusions.

10.2 Checking a tacit assumption: perpendicular lengths

In this section we check a tacit assumption. In deriving the phenomenon of time dilation, we assumed that Bob perceives the height from oven to ceiling to have the same value H that Alice measures it to be. Because Bob and Alice disagree on time intervals, we had better check that assumption.

The length in question is *perpendicular* to Alice and Bob's relative motion. That property is crucial.

Cut two pine boards of equal length – by holding them together while you

Table 10.1. *Pairs of events, as perceived by Alice and Bob, who are in motion relative to each other. The description of each pair runs horizontally. What is surprising is the way Bob's time interval differs from Alice's.*

For Alice		For Bob		Consequence
Spatial separation of events	Time interval between events	Spatial separation of events	Time interval between events	
Not zero	0	Not zero	Not zero!	Relativity of simultaneity
0	Not Zero	Not zero	Not zero – but longer than Alice's!	Time dilation

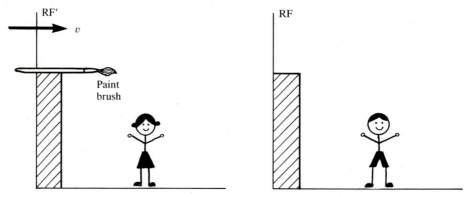

Figure 10.5 Two boards, cut of equal length by holding them together while sawing, are mounted perpendicular to the relative motion of the frames. A paint brush, dipped in vermilion, is affixed to the top of the board in Alice's frame. Note that the relative motion of frames RF and RF′ is just as before, but the origin of Alice's frame has not yet passed the origin of Bob's.

saw them. Place one board in Alice's frame, the other in Bob's, both oriented perpendicular to the motion, as sketched in figure 10.5. Moreover, attach an artist's paint brush to the top of the board in Alice's frame. We examine three possibilities for the length relationship of the boards as perceived by Alice and Bob.

Possibility 1

Try the relationship "the moving board appears to be shorter." The consequences would be the following.

As observed by Bob. The brush would leave a mark on the board in Bob's frame as it passes.

As observed by Alice. Now it is the board in Bob's frame that is seen to move. If "the moving board appears to be shorter," then Bob's board would pass under the brush, and no vermilion streak would be painted on it.

Whether the brush leaves a mark or does not *cannot* depend on the reference frame from which someone observes the relative motion. The suggestion that "the moving board appears to be shorter" leads to a contradiction and must be wrong. (Lest confusion arise later in this chapter, remember that here we are analyzing a length *perpendicular* to the relative motion.)

Possibility 2

Try the relationship "the moving board appears to be longer."

As observed by Bob. The brush would pass over Bob's board.

As observed by Alice. The brush would mark Bob's board as his board passes by.

Again, there is a contradiction; so the suggestion is wrong.

Possibility 3

Try the relationship "the moving board appears to have the same length."

As observed by Bob. The brush grazes the top of Bob's board.

As observed by Alice. The brush grazes the top of Bob's board.

Now we have consistency, and so the suggestion passes the test.

By a process of elimination, our conclusion is this: a length *perpendicular* to the direction of relative motion is measured to be the same in both frames of reference.

Thus our derivation of time dilation is safe. But we are not safe from further surprises; in section 10.5, we will examine lengths *parallel* to the direction of relative motion and find different behavior.

10.3 An interlude on radioactivity

The central relationship in time dilation, equation (1) of section 10.1, contains a square root,

$$\sqrt{1-(v^2/c^2)},$$

a root that recurs frequently in relativity theory. If Alice and Bob are in relative motion at the speed of a jet airplane, 500 miles per hour, say, then the ratio of v to c is

$$\frac{v}{c} = 7.5 \times 10^{-7},$$

a small number. The square root itself works out to be

$$\sqrt{1 - (v^2/c^2)} = 1 - 2.8 \times 10^{-13}$$
$$= 0.999\,999\,999\,999\,72,$$

and that number is extremely close to 1. Alice and Bob would have a hard time noticing time dilation – and that, of course, is why our common sense fails to provide us with any expectation of the effect. It is not part of our daily lives.

Indeed, to observe time dilation one needs extremely high speed or accurate clocks or both. Just as we turned to an atomic particle, the neutral pion, when we needed a swiftly-moving source of light, so we will turn to swiftly-moving radioactive particles for an experiment on time dilation. The radioactive decay provides a kind of timer or clock. To understand how that is possible, we need to learn a bit about radioactive decay itself, and so this interlude arises.

Radioactive decay

An atom of the element uranium has 92 protons in its nucleus and 92 electrons swarming around that positive center. In addition to protons, the nucleus contains *neutrons*, particles quite similar to protons but electrically neutral. The most common form of uranium has 146 neutrons, so that the sum of protons and neutrons equals $92 + 146 = 238$. We focus our attention on a lump of such uranium, the size of a green pea, say.

The uranium nuclei are not stable; they change spontaneously. The dominant mode of change consists of spitting out a tight cluster of two protons and two neutrons; the remaining nucleus has only 90 protons and so is a form of thorium, another metallic element. The lump of material changes slowly over time but remains recognizable as a lump of metallic material.

Figure 10.6 shows the number of uranium nuclei left – the nuclei that have *not* decayed – as a function of time. In the figure, 80 per cent as many nuclei remain at the end of each time interval as were present at the start of the interval. Thus 20 per cent decay in each time interval. Succinctly, in equal time intervals, equal *fractions* of the remaining nuclei decay.

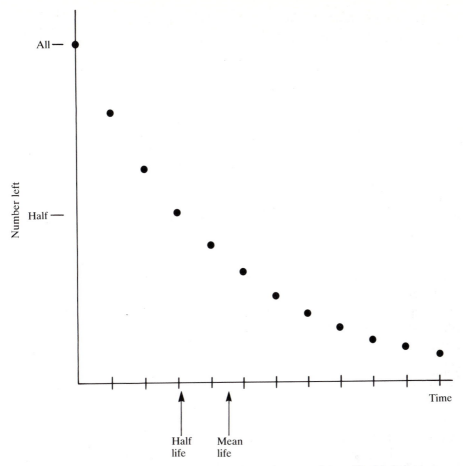

Figure 10.6 The number of uranium nuclei left as a function of time. The labeled arrows indicate the time required for half the nuclei to decay, called the *half life*, and the average lifetime of the nuclei, termed the *mean life*. The time interval indicated by the tick marks was selected so that 80 per cent of the nuclei present at the start of each interval would survive and remain at the end of the interval.

Radioactive uranium nuclei do not "age." Rather, they play a lottery, with odds such that 20 per cent decay in each time interval. Those nuclei that did not decay emerge just as fresh as they were at the very start.

Mean life

The figure shows the time required for half the nuclei to decay and shows also how long a nucleus lives on the average. The average lifetime of the nuclei is usually called the *mean lifetime* or, for short, the *mean life*. Most nuclei, as they play the lottery, decay before time has advanced to the value of the mean life. But some nuclei live to twice the mean life or three times it

Table 10.2. *Some radioactive particles and the mean life of each. The integer immediately following an element name gives the sum of the protons and neutrons in the nucleus. The element name and that integer uniquely identify the type of nucleus.*

Particle	Mean life
Uranium-238	6.5×10^9 years
Thorium-234	35 days
Protactinium-234	9.6 hours
Uranium-234	3.6×10^5 years
Thorium-230	1.1×10^5 years
Radium-226	2.3×10^3 years
Radon-222	5.5 days
Polonium-218	4.4 minutes
Lead-214	39 minutes
Bismuth-214	28 minutes
Polonium-214	2.4×10^{-4} seconds
Lead-210	32 years
Bismuth-210	7.2 days
Polonium-210	200 days
Lead-206	stable
Carbon-14	8300 years
Tritium (hydrogen-3)	18 years
Neutral pion	8.2×10^{-17} seconds
Charged pion	2.6×10^{-8} seconds
Muon	2.2×10^{-6} seconds

or even more. Their great age at the instant of decay makes up for the smaller number of nuclei with ages exceeding the mean life. Thus their longevity puts the mean life out where it lies relative to the time in which half the nuclei decay.

For our form of uranium, the mean life is enormous, 6.5×10^9 years, and that explains why such uranium is still found in the earth's crust.

Table 10.2 lists the mean life for a variety of radioactive particles. The columns start with our form of uranium and go through a sequence of decays to radon, a radioactive gas present in the earth. The percolation of radon from the ground into houses emerged in the 1980s as a newly-found

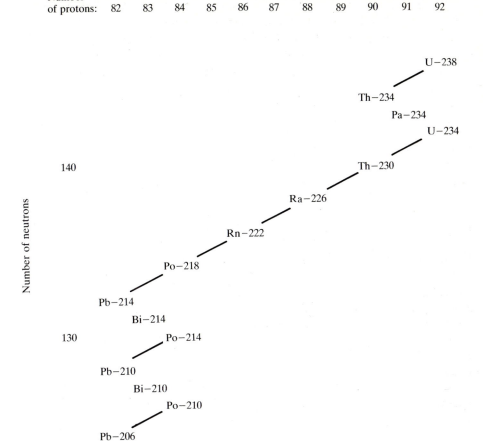

Figure 10.7 The decay sequence: uranium-238 to radon and thence to lead-206. In the perhaps-puzzling abbreviation "Pb," the letters denote "lead," from the Latin "plumbum." In ancient Rome, a "plumber" repaired lead pipes. In some of the decay steps, the sum of the protons and neutrons does not change. A neutron merely changes into a proton, and an electron is emitted.

health hazard. Radon decays with a mean life of 5.5 days, initiating a decay sequence that ends with a stable form of lead. Figure 10.7 shows the entire sequence of decays, from uranium through radon to stable lead.

In figure 10.7, each element appears only in the column corresponding to its number of protons, for example, 92 for uranium. Different forms of uranium with different numbers of neutrons, uranium-238 and uranium-234, are called different *isotopes* of uranium. The word "isotope" comes from the Greek roots "*isos*," meaning "the same," and "*topos*," meaning "place." Isotopes of uranium appear in "the same place" in the periodic table of elements, namely, the box for all forms of uranium: 92 protons and any number of neutrons.

The decay chain from uranium-238 passes through two unstable isotopes of lead before ending in a stable isotope, lead-206. Along the way, the chain passes through two isotopes of thorium, two of bismuth, and three of polonium. It is quite natural for an element to exist in various isotopic forms, but many of the isotopes will be radioactively unstable. (Indeed, sometimes all of them are unstable.)

Table 10.2 lists two radioactive isotopes of carbon and hydrogen. Both carbon-14 and tritium act chemically like the ordinary stable forms of carbon and hydrogen, respectively. When the radioactive isotopes are used in biological research, biochemical reactions incorporate those isotopes into molecules just as the reactions would incorporate ordinary carbon and hydrogen. The radioactive forms can be detected through their decay products, and so those forms serve as tracers of the metabolic paths taken by carbon and hydrogen.

Since the late 1940s, carbon-14 has been used to date wooden and other organic archeological finds. While a tree lives, normal metabolism and exchange with the environment keep its percentage of carbon-14 (relative to all isotopes of carbon) the same as that in nature at large. After the tree has died and become charcoal in the Lascaux cave of France, say, the carbon-14 continues to decay but is no longer replaced. Thus the percentage of carbon-14 decreases with time. A measurement of the relative amount of carbon-14 tells an archeologist when the tree fell and the cave paintings were drawn.

In chapter 8, the neutral pion served as a swiftly-moving light source. It decays into two gamma rays with an extremely short mean life. Electrically charged pions decay more slowly. Their decay produces a particle that is new to us: the muon. A negatively charged muon is about 200 times as massive as an electron and has the same amount of charge. It is unstable and decays with a mean life of 2.2×10^{-6} seconds. Muons play a major role in the next section.

Synopsis

A capsule summary of this section is in order. Radioactive decay is a common occurrence, both in nature and in the laboratory. Decay is a statistical process, a kind of lottery that the particle plays, the outcomes being survival or decay. Nonetheless, the average lifetime of a million uranium-238 nuclei or a million muons is a well-defined quantity and is reproducible. Another million uranium-238 nuclei or muons will exhibit the same average lifetime (with insignificant statistical fluctuations).

What we have *not* addressed in this section is *how* radioactive decay occurs, nor have we even listed all possible modes of decay. Those are fascinating topics. Alas, to develop them would take us too far afield in what is already just an interlude, something to prepare us for an actual experiment on time dilation. Chapter 11 will say more about radioactive decay; for now, we return to time dilation.

10.4 A test of time dilation

High in the earth's atmosphere, muons are regularly produced in a thoroughly natural way. High-speed protons (originating outside the solar system but within our galaxy) collide with oxygen and nitrogen nuclei in the upper reaches of the atmosphere, producing charged pions. The pions quickly decay, producing muons. The muons form a novel kind of rain coming down through the lower reaches of the atmosphere. The experiment that we study, an experiment performed by David H. Frisch and James H. Smith in the early 1960s, uses that rain of muons.

Figure 10.8 sketches the essential locations in the experiment. Muons are born high in the atmosphere. They are first observed at the top of Mount Washington in the state of New Hampshire, about 2000 meters above sea level. That constitutes event 1. Subsequently, the muons are observed at sea level; that constitutes event 2. The number of muons observed (per hour) at sea level is about 70 per cent of the number observed on the mountain top. Evidently some muons decay between the mountain elevation and sea level, thus reducing the rate of "rain."

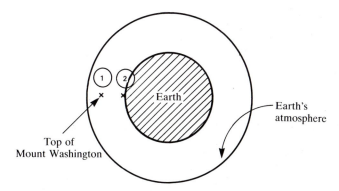

Figure 10.8 The essential locations in the descent of the muons.

The two physicists selected muons traveling at the high speed of $v=0.995c$. (In detail, Frisch and Smith selected muons of specific energy and used an experimentally-tested relationship between speed and energy. Thus $0.995c$ is an experimental number and beyond reasonable doubt.) Placing ourselves in the role of the physicists, we calculate the time (as noted by us) that the muons take to travel from the mountain top to sea level. We reason that

$$v \times \left(\begin{array}{c}\text{elapsed time}\\ \text{for us}\end{array}\right) = \left(\begin{array}{c}\text{height of Mount Washington}\\ \text{for us}\end{array}\right),$$

and so

$$\left(\begin{array}{c}\text{elapsed time}\\ \text{for us}\end{array}\right) = \frac{2000 \text{ meters}}{0.995 \times 3 \times 10^8 \text{ meters}/\text{seconds}}$$

$$= 6.7 \times 10^{-6} \text{ seconds}.$$

Table 10.2 told us that the mean life of a muon is 2.2×10^{-6} seconds. Thus the typical muon lives only some 2×10^{-6} seconds, about $\frac{1}{3}$ of the travel time we have computed. How can 70 per cent of the muons make a trip to sea level that takes about three times as long as the typical muon lives?

Figure 10.9 provides for muons a graph similar to figure 10.6 for uranium-238. After 6.7×10^{-6} seconds have elapsed, only 5 per cent of the muons are expected to have survived. That is a far cry from 70 per cent.

To resolve the paradox, we need to look at life from the muon's point of view. A muon is at rest in a reference frame moving with it. When seen from that reference frame, the two "observation" events, events 1 and 2, occur at the same location: the muon's fixed position in that frame. The general time dilation relationship.

$$\Delta t_{\text{any other frame with relative speed } v} = \frac{1}{\sqrt{1-(v^2/c^2)}} \Delta t_{\text{frame where events occur at } the \ same \ place},$$

becomes

$$\Delta t_{\text{observed by us}} = \frac{1}{\sqrt{1-(v^2/c^2)}} \Delta t_{\text{muon's frame}}.$$

To solve for the elapsed time as the muon perceives it, we multiply both sides by the square root, finding

$$\Delta t_{\text{muon's frame}} = \sqrt{1-(v^2/c^2)} \ \Delta t_{\text{observed by us}}.$$

Because the square root is less than 1, the muon perceives a shorter time. (The muon takes Alice's place, and we take Bob's place.) This is a step in

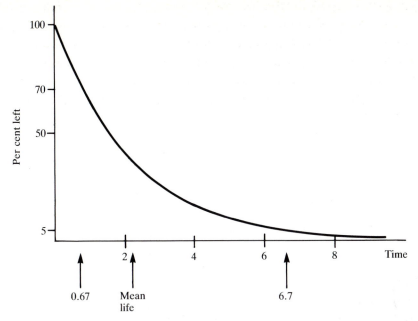

Figure 10.9 For a collection of muons at rest in the laboratory, the graph shows the percentage remaining (that is, having *not* decayed) as a function of time. The time axis is labeled in units of 10^{-6} seconds. Thus "2" means 2×10^{-6} seconds.

the right direction: because the muon perceives a shorter time, the muon is more likely to have survived and not yet decayed.

How much shorter is the perceived travel time? Substituting $v=0.995c$, we get

$$\Delta t_{\text{muon's frame}} = 0.1 \; \Delta t_{\text{observed by us}}$$

$$= 0.1 \times 6.7 \times 10^{-6}$$

$$= 0.67 \times 10^{-6} \text{ seconds.}$$

As far as a muon is concerned, the elapsed time is smaller by a factor of $\frac{1}{10}$. The elapsed time is only 0.67×10^{-6} seconds, to be compared with a mean life of 2.2×10^{-6} seconds. In the muon's frame, only a fraction of a mean life has elapsed, specifically, the fraction $0.67/2.2 \approx \frac{1}{3}$. So, of course, most of the muons have not yet decayed. Indeed, inspection of figure 10.9 shows that about 70 per cent of the muons should remain, just what Frisch and Smith found.

Without the idea of time dilation, the experimental results would remain paradoxical. Once we introduce time dilation, everything falls into place.

The decay of muons in high-speed flight (relative to us) provides excellent support for the special theory of relativity.

Another remark is in order: about the way the experiment was done. The observations, events 1 and 2, were not made on the very same muons. The "rain rate" of muons (with speed $v=0.995c$) was first measured at the top of Mount Washington. Then the apparatus was moved to sea level (at Cambridge, Massachusetts), and the rain rate was measured again. At any fixed location on the earth's surface, the rain rate of muons is substantially constant in time, and it is also unaffected by small shifts in latitude and longitude. One should think of a nice, steady, all-day rain, one that pervades the countryside. You can make your measurements any time and any place you wish – but the rain rate is lower near the ground because some of the drops evaporate (or "decay") before they reach the ground.

To continue the analogy, let us suppose that you are standing on the porch, watching the rain fall past the top of the flowering cherry and into a puddle on the driveway. You note a certain time for a raindrop to descend from tree top to puddle. For the raindrop itself (or an observer riding with the raindrop), the events of passing the tree top and entering the puddle happen at the same location, namely, at the raindrop itself. For the rain-drop (or observer), the time between those events is less than it is for you. The "why" of this, of course, goes back to the diagrams for Alice and Bob in figures 10.1 and 10.2 plus the remarkable constancy of the speed of light.

10.5 Length contraction

The last paragraph of section 10.2 promised that we would examine lengths parallel to the direction of relative motion. Is such a length a relative notion, relative to the reference frame from which the object is observed?

To find out, we return to Alice and Bob, and we suppose that Alice has just caught a brook trout of record size. Precisely how long is that trout?

Once subdued, the fish is at rest in Alice's frame. Alice measures its length to be ℓ_0, say. The subscript zero (on the ℓ for "length") denotes zero speed, that is, ℓ_0 is the trout's length when measured in a frame where the fish is at rest (and has zero speed). Such a reference frame is called the *rest frame* of the fish.

Alice and her trout move with speed v relative to Bob. If we denote by unadorned ℓ the trout's length as Bob measures it, the central question becomes this: how are ℓ and ℓ_0 related?

To answer the question, we need to analyze the ways in which Alice and

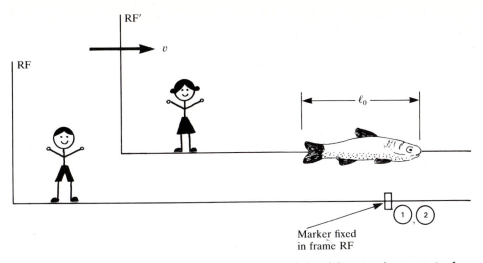

Figure 10.10 How Alice and Bob can measure the trout's length by using the same pair of events. The trout is at rest in Alice's frame, and the marker is fixed in Bob's frame.

Event 1: Fish's nose passes the marker.

Event 2: Tail passes the marker.

Only so that we can see things more clearly does the sketch show Alice's frame displaced upward from Bob's; continue to think of Alice's frame as moving along Bob's frame just as it has in the past.

Bob might actually determine the lengths ℓ_0 and ℓ. Figure 10.10 sketches the essentials of a good method. Here are the steps.

Step (a). Bob notes how long it takes the trout to pass a marker fixed in his frame; let us call that elapsed time Δt_{Bob}. To determine the length ℓ, Bob multiplies the elapsed time by the trout's speed v:

$$\ell = v\Delta t_{\text{Bob}}.$$

Step (b). As Alice observes things from her frame, it is the marker that moves, not the trout. The marker moves from the trout's head to its tail, a distance ℓ_0 (for Alice). Alice denotes the elapsed time by Δt_{Alice}. Thus Alice writes

$$\ell_0 = v\Delta t_{\text{Alice}}.$$

Step (c). Now we recall the general time dilation relationship:

$$\Delta t_{\text{any other frame with relative speed } v} = \frac{1}{\sqrt{1-(v^2/c^2)}} \, \Delta t_{\text{frame where events occur at } the\ same\ place}.$$

In the fish-measuring context, the two events happen at the same place in Bob's frame, that is, at the marker fixed in his frame; so the relationship of time intervals is

$$\Delta t_{\text{Alice}} = \frac{1}{\sqrt{1-(v^2/c^2)}} \ \Delta t_{\text{Bob}}.$$

We solve for Δt_{Bob} (by multiplying both sides by the square root) and then substitute for Δt_{Bob} in the equation of step (*a*):

$$\ell = v \ \sqrt{1-(v^2/c^2)} \ \Delta t_{\text{Alice}}.$$

Step (*b*) told us that ℓ_0 equals the product $v \Delta t_{\text{Alice}}$; so we substitute on the right-hand side:

$$\ell = \sqrt{1-(v^2/c^2)} \ \ell_0.$$

Because the square root is less than 1, Bob measures a smaller length for the trout than Alice does. This effect is called *length contraction*.

Let me display our result as follows: the equation

$$\ell = \sqrt{1-(v^2/c^2)} \ \ell_0 \tag{1}$$

means

$$\begin{pmatrix} \text{length as measured in frame} \\ \text{where object has speed } v \end{pmatrix} = \sqrt{1-(v^2/c^2)} \times$$

$$\begin{pmatrix} \text{length as measured in frame where} \\ \text{object has ``zero speed'': its rest frame} \end{pmatrix}.$$

This relationship pertains to length *parallel* to the direction of relative motion. (As we worked out in section 10.2, lengths *perpendicular* to the direction of relative motion are measured to be the same in both frames of reference.)

Conclusion

The general, detailed statement in equation (1) is the fruit of our analysis. A quick verbal summary is this:

> Because the square root in equation (1) is less than 1, an object has its greatest length when measured in its rest frame.

If we look back over our derivation, we see that there are only two essential elements: (1) the idea that distance equals speed times elapsed time, and (2) a comparison of elapsed times, Bob's and Alice's. The first item is neither novel nor controversial. The comparison of elapsed times, however, brings in time dilation and so leads to the unexpected length contraction (as Bob perceives the trout). In turn, time dilation arises from the constancy of the speed of light. Thus we may say that length contraction is another consequence of that remarkable constancy.

10.6 The muons revisited

In section 10.4, we met the paradox of the muons: how can some 70 per cent of the muons survive the trip from Mount Washington to sea level? Time dilation provided a resolution: for the muons, the time interval is only $\frac{1}{10}$ as long as for us.

Length contraction provides another, equivalent way to resolve the paradox. Mount Washington is at rest in our frame of reference, and we measure it – from sea level to peak – to be 2000 meters high.

From the muon's point of view, it is the mountain that moves – and at speed $v=0.995c$. First the top passes; then the slopes rush by; and soon sea level arrives at the muon. How high, for the muon, is the mountain? Equation (1) of the last section responds with

$$\begin{pmatrix} \text{Height measured} \\ \text{by muon} \end{pmatrix} = \sqrt{1-(v^2/c^2)} \times \begin{pmatrix} \text{height measured} \\ \text{by us} \end{pmatrix}$$

$$= 0.1 \times 2000 \text{ meters}$$

$$= 200 \text{ meters.}$$

How shrunken!

Only a hill 200 meters high need pass the muon. Moreover, the hill is rushing at the muon with speed $0.995c$. Once the top passes, sea level reaches the muon in the time

$$\Delta t_{\text{muon's frame}} = \frac{\text{height observed by muon}}{\text{speed of hill}}$$

$$= \frac{200 \text{ meters}}{0.995 \times 3 \times 10^8 \text{ meters}/\text{second}}$$

$$= 0.67 \times 10^{-6} \text{ seconds.}$$

That time is substantially less than the mean life of a muon (which is 2.2×10^{-6} seconds), and so most of the muons are still around when sea level arrives. Indeed, the value for the elapsed time as noted in the muon's frame is the same here as it was when we used time dilation to resolve the paradox. We have consistency.

10.7 Compendium

A survey of what we have achieved to date in relativity theory is in order.

The Principles

The special theory of relativity is founded on two principles.

Principle 1. The Principle of Relativity, stated in either of two equivalent forms:

> Colloquial statement: if we are in an unaccelerated vehicle, its motion has no effect on the way things happen inside it.

> Formal statement: the laws of physics are the same in all unaccelerated reference frames.

Principle 2. The motion of light is not affected by the motion of the source of light.

The principles are a generalization from observations and experiments. For us, the 1964 experiment on light from swiftly-moving neutral pions provides the reason to assert Principle 2.

Principle 1 is appropriately called the Principle of Relativity, but the reason is not patently obvious. Principle 1 implies that there is no absolute standard for uniform motion. In the figures of this chapter, Alice is drawn as moving (relative to Bob). But Alice and Bob observe the same laws of physics in their respective reference frames; there is no intrinsic difference. So when Alice opines that Bob is moving (relative to her), she is fully as justified as is Bob when he notes that Alice moves (relative to him). Uniform motion is meaningful only relative to some chosen reference frame (Alice's or Bob's or any of a million others).

Here is another way to put it. As you wake up from your nap, you hear a voice saying ". . . is 450 miles per hour. And that's all for now, folks." You look around. Yes, you are in the airplane, not in the terminal, and so that was the captain's voice over the intercom. But are you approaching Phoenix at 450 miles per hour? Maybe – but maybe not. If the pilot was quoting the plane's speed relative to the ground, called the *ground speed*, then the answer is yes. But the captain may have been citing the plane's speed relative to the ambient air, called the *air speed*. (The air speed determines the drag on the aircraft and also the lift provided by its wings. A pilot conscientiously monitors both the ground speed and the air speed.) If the weather is moving air toward Phoenix at 150 miles per hour and thus is providing a tail wind of 150 miles per hour and if your air speed is 450 miles per hour, then you are actually approaching Phoenix at about 600 miles per hour. A mere statement that the plane's speed is 450 miles per hour is ambiguous. Relative to what – ground or air – is the plane going 450 miles per hour? A numerical speed without a reference frame – given explicitly or implicitly – is not a meaningful declaration.

In short, both forms of Principle 1 imply that uniform motion is always a relative concept: uniform motion is meaningful only relative to a reference frame, and the frame may be chosen at will. In this sense, both forms of Principle 1 provide a "principle of relativity" for uniform motion.

The constancy of the speed of light

Observers in all unaccelerated reference frames measure the same speed for light (in vacuum) from any given source. That common speed is always c, where

$$c = 2.997925 \times 10^8 \text{ meters / second}$$

$$\approx 3 \times 10^8 \text{ meters / second}.$$

In short, they *all* measure 3×10^8 meters/second *always* for light in vacuum. This remarkable property is called the *constancy of the speed of light*. We deduced this result by using Principle 1 and Principle 2 together.

The relativity of simultaneity

From the constancy of the speed of light we deduced that

> Spatially-separated events that are simultaneous in one frame are, in general, not simultaneous when viewed from another frame.

An absolute, global "now" does not exist. Rather, simultaneity is a relative notion, relative to the reference frame from which two events are observed.

Time dilation

The constancy of the speed of light led us (via the Pythagorean theorem) to the conclusion that

$$\Delta t_{\text{any other frame with relative speed } v} = \frac{1}{\sqrt{1-(v^2/c^2)}} \, \Delta t_{\text{frame where events occur at } \textit{the same place}}.$$

The time interval between a pair of events is least when observed in a frame where the events occur at the same place. For other observers, the time interval is longer: it is "dilated."

The decay of muons in high-speed flight (relative to the earth) provides fine experimental support for this prediction.

We find that a time interval is a relative notion, relative to the reference frame from which the events at the start and end of the interval are observed. The relativity of simultaneity says the same thing, but for a pair of events that are simultaneous for some observer.

Perpendicular lengths

A length perpendicular to the direction of relative motion is measured to be the same in both frames of reference.

Parallel lengths

For the length of an actual object, such as Alice's trout, we deduced that

$$\ell = \sqrt{1 - (v^2 / c^2)} \; \ell_0,$$

which means

$$\left(\begin{array}{c} \text{length as measured in frame} \\ \text{where object has speed } v \end{array} \right) = \sqrt{1 - (v^2 / c^2)} \times$$

$$\left(\begin{array}{c} \text{length as measured in frame where} \\ \text{object has ``zero speed'': its rest frame} \end{array} \right)$$

This relationship pertains to length parallel to the direction of relative motion. It carries the name "length contraction" because an observer for whom the object is moving measures a shorter length than an observer for whom the object is at rest. (Neither observer is making an error. Rather, each observer is correct, but our intuition needs revision.)

The phenomenon of length contraction is supported experimentally by the decay of muons in high-speed flight (relative to the earth).

Lengths form the basis for discussing space. Because the length of an object is a relative notion, relative to the reference frame from which one measures the object, we find that space itself has a relative aspect. Space retains three dimensions for all observers, but the extent of an object and hence of space depend on the reference frame.

Additional resources

David Frisch and James Smith describe their muon experiment in the *American Journal of Physics*, volume 31, pages 342–55 (1963). They did the experiment to produce a film, *Time Dilation: An Experiment with Mu-mesons*, part of the PSSC series and available from the same source as John King's *Photons*, cited in "Additional resources" for chapter 7. The muon film is well worth a class period's time.

The film establishes the reality of time dilation and also shows the intricacy of a professional experiment.

Questions

1. A firecracker goes off in Houston, Texas. A time 0.03 seconds later (as measured on synchronized earth clocks), another firecracker goes off in Great Falls, Montana, 2400 kilometers away (as measured on earth).

(a) Draw a sketch of the context. It will help you to picture the questions.

(b) How fast must a rocket ship travel if it is to be present at both events?

(c) What will the rocket ship pilot note for the time interval between the events? (Be careful to express all speeds in the same units; here, meters/second is preferable to kilometers/second. Recall that the prefix "kilo" means "thousand.")

(d) What is the numerical difference in the two times, that measured on earth versus that measured in the rocket ship? Is the difference consistent with the verbal, qualitative statement made at the start of the "Comments" in section 10.1?

2. Charlie has just caught a lake trout 20 inches long. Zipping by in her motor boat, the game warden sees the fish as 12 inches long. Uh-oh! the minimum legal length is 16 inches.

(a) How fast was the game warden going? (Note. To solve an equation of the form

$$a = \sqrt{1 - x^2}$$

for x, first square both sides. Then isolate x^2 on one side of the equation and take a square root.)

(b) Will Charlie have to pay a fine? Briefly, why?

3. Firecrackers are placed at points A and B, which are 100 meters apart as measured in the rest frame of the earth. As a rocket ship, moving at speed $v = \frac{3}{5}c$, passes point A, the firecracker there explodes. As the rocket passes B, the second firecracker explodes. According to a clock on the rocket ship, the second explosion occurs a time $\Delta t_{\text{rocket's frame}}$ after the first. Clocks synchronized in the rest frame of the earth measure a time interval $\Delta t_{\text{earth's-frame}}$ between the explosions.

(a) Determine numerically $\Delta t_{\text{earth's frame}}$.

(b) Determine $\Delta t_{\text{rocket's frame}}$ by reasoning with time dilation.

(c) Determine $\Delta t_{\text{rocket's frame}}$ by reasoning with length contraction.

(d) Do your results in parts (b) and (c) agree? Should they?

(e) To what situation that you have met before is this example analogous?

4. Recall the neutral pions that decay into two gamma rays. In its own frame of reference, a neutral pion lives only 8.2×10^{-17} seconds (on the average). In the experiment described in chapter 8, the pions moved relative to the laboratory with a speed within 1 per cent of the speed of light. Indirect evidence implies a speed of $0.99975c$. (Note that the neutral pions are *different* from the muons whose journey from Mount Washington to sea level we discussed in this chapter.)

(a) What would be the average lifetime of the neutral pions if the lifetime were measured in the laboratory (where the pions are in motion)?

(b) How far can the pions travel in that time? Convert your answer to millimeters.

(c) Can you see why indirect means must be used to study neutral pions? Use a sentence to explain.

5. What is the fastest-moving vehicle you have ever seen? How long does it take to make a typical trip (as noted by clocks on earth)? Now that you know about relativity theory, calculate the duration that a passenger aboard the vehicle would note the trip to take. How much shorter or longer does the trip take for the passenger?

If you use a hand calculator and it gives you sensible results, then you are all set. If not, then here is an approximation that you may find useful:

$$\sqrt{1-(v^2/c^2)} \approx 1 - \tfrac{1}{2}\frac{v^2}{c^2} \quad \text{provided } \frac{v}{c} \ll 1.$$

You can prove this relationship by squaring both sides and reasoning that the "extra" term on one side is negligible because of the strong inequality, $(v/c) \ll 1$. If you do use the approximation, then you will probably find it useful to "distribute" your numbers, recalling that

$$(1-a)\,b = b - ab$$

for any constants a and b.

6. On the flat Nevada desert, two firecrackres go off 10^{-4} seconds apart at locations separated by 2.4×10^4 meters, as noted by the local residents.

(a) Draw a sketch of this context.

(b) How fast would a race car need to travel so that both events occur at the same position relative to the race car, for example, right next to the driver's door?

(c) What would be the time interval between the two events as perceived by the driver?

7. Point out the error in the following argument. A pencil, at rest in Bob's frame, has a length $\ell_{\text{pencil, Bob's frame}}$ as measured in that frame. With respect to Alice's frame, the pencil is moving, and Alice measures it to have a shorter length, $\ell_{\text{pencil, Alice's frame}}$. Therefore it is possible to distinguish between two unaccelerated reference frames, and the Principle of Relativity is not valid. (Note. Alice and Bob move relative to each other with constant speed v, and both of them inhabit unaccelerated reference frames, just as they have throughout chapters 9 and 10.)

Bear in mind that objects are not invariably at rest in Alice's frame or anyone else's frame. The trout happened to be at rest in Alice's frame in our derivation of length contraction. In each context, you have to look for the intrinsically significant features. What's at rest in which frame? In which frame (if in any) do the events occur at the same place?

You may find it useful to introduce an auxiliary pencil.

8. The length from beak to tail of a peregrine falcon is typically 48 centimeters, as measured by an ornithologist inspecting a specimen in the museum. When plunging after prey, peregrines reach speeds of 125 miles per hour (about 56 meters/second). How fast would a peregrine need to fly so that a bird watcher on the ground would observe its length to be one-half of the typical value? Does even a peregrine come close to that speed?

9. Alice and Bob have their usual relative motion; moreover, each wants to cook a soft-boiled egg. As their stoves pass each other, both Alice and Bob put an egg into boiling water and set timers for 5 minutes.

For Alice, whose egg is ready first, hers or Bob's? And for Bob, whose egg is ready first? Some sketches will help you to puzzle this out. Describe whatever consistency you find, and explain why it arises.

10. In what way does the *speed* of light play an essential role in our derivation of time dilation?

11 $E=mc^2$

Does the inertia of a body depend on its energy content?

Albert Einstein,
title of his first paper on $E=mc^2$

11.1 Energy, mass, and momentum reviewed

Easily the most famous result of relativity theory is the equation $E=mc^2$. In this chapter, we learn where the equation comes from, what it means, and what it does not mean. To do so, we need to be clear about what certain terms in physics mean, and so this section reviews the essential concepts. If you have not yet read appendix A, "Energy," or have not done so recently, please read it before proceeding any farther in this chapter. Then this section will indeed be a review.

Because this section is a review, the topics are presented in telegraphic style.

> *Energy:* the ability to do work, for example, to lift a weight. Energy is an attribute of a physical object or of whatever is contained in a specified region of space.

As figure 11.1 shows, a moving object can be used to lift a weight. Thus there is energy associated with motion, called "kinetic" energy, from the Greek verb "*kinein*," meaning "to move."

The higher the object starts, the more kinetic energy it has when it hits. Thus the greater the height, the more potential for kinetic energy at the bottom. Physics introduces the idea of a "gravitational potential energy," an energy associated with position in the earth's gravitational field.

The test for any kind of potential energy is this: can we convert it to kinetic energy (and then surely lift a weight)?

Table 11.1 lists various forms of energy. Electrical potential energy arises from the electrical attraction or repulsion of electrically charged particles, such as electrons and protons. Similarly, nuclear potential energy arises from the specifically nuclear forces among protons and neutrons, forces that keep those particles bound together in an atomic nucleus. After all, the protons repel one another electrically and tend to disrupt the nucleus.

Table 11.1. *Various forms of energy.*

Kinetic	Potential	Radiant
The energy associated with motion	Energy that has the potential for being converted to kinetic energy	The energy of electromagnetic waves (or, equivalently, of photons)
	Examples: gravitational electrical nuclear	

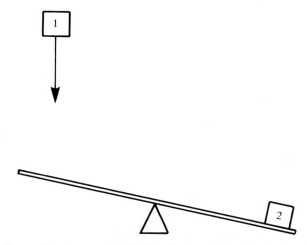

Figure 11.1 Drop object 1 onto the left side of a pivoted board, and object 2 (a weight) is lifted into the air.

There must be some other force, largely attractive, that holds the protons and neutrons together. Indeed, there is, and one calls it the nuclear force.

There is a law of conservation of energy: energy, when considered in all its forms, is automatically conserved. That is, the total amount of energy remains numerically constant in time (although some of the energy may swap from one form to another). Whenever one object exerts a force on another object – either a push or a pull – physics can introduce a potential energy (or other form of energy) such that a law of conservation of energy holds. This is, of course, a remarkable property of nature, not just a sign of physicists' ingenuity.

Mass: inertia; sluggishness; the inherent reluctance to undergo a change in velocity. Remember always that mass is an attribute, not a thing. Mass is the attribute "inertia" of a physical object or of whatever is contained in a specified region of space.

The word "mass" has had other meanings in English, and some of them persist. Fading from use is the meaning "quantity of matter." Whatever historical value this meaning may have, the meaning is *not* the sense in which this book uses the word "mass." In the physics of this chapter, "mass" means inertia, which is an attribute, not a thing. If one consults the research documents of the decade 1900–10, when Einstein and others developed the equation $E=mc^2$, one finds that – for those physicists – the word "mass" was a synonym for "inertia." In this book, " 'mass' means inertia" is the appropriate usage.

Matter: tangible stuff; what you can hold in your hand.

Note that all matter has inertia and hence has mass. Something that has mass, however, need not necessarily be made of matter. Thus – and note this well – the concepts "mass" and "matter" are distinct.

Momentum: the product "mass times velocity."

A law of conservation of momentum is more complicated than energy conservation because momentum has a direction. We can state the law this way: if no forces from outside a physical system act in a certain direction, then the amount of momentum in that direction is automatically conserved.

We need to be clear about a certain distinction. Matter, mass, and energy: in what sense do they *exist*? Take a dime out of your pocket and flip it into the air. The dime is an example of matter; matter, we see, exists as a *thing*. Traveling through the air, the dime has some inertia – it is not blown around by every little breeze – and it has energy, partly kinetic energy, partly gravitational potential energy. Thus mass and energy exist as *attributes* of a physical system, just as do color or shape. Mass and energy exist in the same fashion, and that fashion is different from the way matter exists. Bearing this distinction in mind will spare us much grief later on.

11.2 A thought experiment for $\Delta E=(\Delta m)c^2$

With review accomplished, we turn now to an "experiment" that is the crucial step in this chapter. The word "experiment" appears in quotation marks because we merely think about a possible experiment, rather than discuss an experiment that has been published. In the language of physics, we do a "thought" experiment, using deductive logic in place of actual

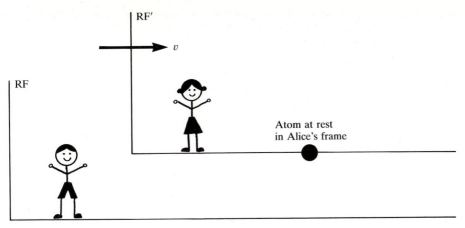

Figure 11.2 The context for our thought experiment. Alice's frame moves with speed v relative to Bob's frame.

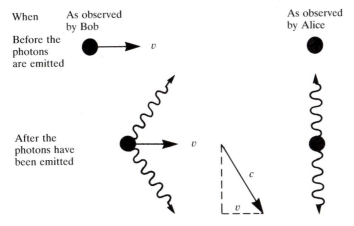

Figure 11.3 How Alice and Bob observe the process. The triangle shows how the diagonally downward light must travel, as perceived by Bob.

apparatus. A thought experiment is an aid to theoretical reasoning. It is not meant to be a substitute for an actual experiment or to masquerade as one. Rather, it has virtues of its own.

Figure 11.2 shows Alice, Bob, and an atom at rest in Alice's frame. The atom emits two photons of equal energy, and we specify that, in Alice's frame, the two photons travel back-to-back and perpendicular to the direction of the frames' relative motion. Our strategy is to compare the ways Alice and Bob describe this process and to extract a major result from the comparison.

Figure 11.3 depicts what Alice and Bob observe. Before the photons are emitted, the atom is at rest in Alice's frame and moves rightward with speed *v* as observed from Bob's frame.

In section 6.3, we learned that a photon has a momentum of magnitude *hf/c*; the direction of the momentum is simply the direction in which the light is traveling. Momentum plays a central role as we analyze the emission process, first from Alice's point of view and then from Bob's.

In Alice's frame, the momenta of the two photons are equal in magnitude but opposite in direction. The sum of their momenta is zero. The atom had no momentum to start with (as perceived in Alice's frame); the photons, taken together, have no net momentum; and so conservation of momentum says the atom must have no momentum afterward. The atom must remain at rest. (Symmetry, too, requires that conclusion. How could the atom choose a direction in which to recoil? Being unable to choose, the atom just sits tight.)

As observed from Bob's frame, the situation is significantly different. The photons move along diagonal directions. The photons must keep up with the atom's forward motion (because, as observed by Alice, the two photons and the atom lie along a straight line), and so each photon's component of velocity parallel to the atom's velocity must be *v*. The triangle in the figure shows this.

Thus the photons' momentum parallel to the atom's velocity is this:

$$\left(\begin{array}{c}\text{Number of}\\\text{photons}\end{array}\right) \times \left(\begin{array}{c}\text{entire momentum}\\\text{of one photon}\end{array}\right) \times \left(\begin{array}{c}\text{fraction that}\\\text{is parallel}\end{array}\right) = 2\,\frac{hf}{c}\,\frac{v}{c}.$$

The fraction of each momentum that is parallel to the atom's velocity is the ratio of two sides of the triangle, the side *v* over the hypotenuse *c*.

Now we examine energy and momentum changes for the atom as observed by Bob. Because the atom emits two photons, each of energy *hf*, energy conservation implies that the atom's energy decreases:

$$\Delta \text{ Energy}_{\text{atom}} = -\,2hf.$$

Recall that capital Greek delta, Δ, means "change in" for whatever follows the letter.

Conservation of momentum implies that the *change* in the atom's momentum must be equal in magnitude to the momentum of the photons but opposite in direction. Thus

$$\Delta \text{ momentum}_{\text{atom}} = -\,2\frac{hf}{c}\frac{v}{c}$$

$$= \Delta \text{ Energy}_{\text{atom}}\,\frac{v}{c^2}.$$

The second step follows by substitution for $-2hf$ from above.

The definition of momentum (as mass times velocity) implies

$$\Delta \text{ momentum}_{\text{atom}} = (\Delta \text{ mass}_{\text{atom}}) \ v$$

because the atom's velocity remains constant here. (Because the atom remains at rest as perceived by Alice, the atom continues to move with constant speed v as observed by Bob.)

The two expressions for the change in the atom's momentum must agree, and so we deduce that

$$\Delta \text{ Energy}_{\text{atom}} \frac{v}{c^2} = (\Delta \text{ mass}_{\text{atom}}) \ v.$$

Multiplication on both sides by c^2 and division by v produces the equation

$$\Delta \text{ Energy}_{\text{atom}} = (\Delta \text{ mass}_{\text{atom}}) \ c^2.$$

In short,

$$\Delta E = (\Delta m) \, c^2,$$

where E and m denote the atom's energy and mass, respectively. In words, when the energy of the atom decreased, so did its mass.

Note that the atom has just as many electrons, protons, and neutrons as before. Only the electrical potential energy and the kinetic energy of the electrons have changed. (The emission of light – that is, the emission of the two photons – is accompanied by a change in how the electrons swarm around the nucleus, and that change alters the electrical potential energy and the kinetic energy.) The change in the atom's energy is accompanied, however, by a change in the atom's inertia. When the energy decreased, so did the inertia.

We have now a connection between ΔE and Δm, between changes in energy and in mass. We have not yet arrived at $E=mc^2$, but we are close.

Synopsis

Here is a summary of the reasoning, presented strictly in verbal form in two paragraphs.

In Alice's frame, the photons (of equal energy) are emitted back-to-back, and so the atom remains at rest in her frame. Because the atom's state of motion does not change in Alice's frame, it does not change in Bob's frame either.

In Bob's frame, the atom moves with speed v. The travel directions of the photons are tilted toward the direction of the atom's velocity. Thus the photons' momenta, when added together, give a positive component along the direction of the atom's velocity. In the isolated system of atom and photons, momentum (as well as energy) must be conserved, that is, must remain constant in time. In Bob's frame, the photons contribute a positive amount of momentum that was not present initially. To compensate, the atom's momentum must decrease. Because momentum is the product of mass and velocity and because the atom's velocity does not change, we conclude that the atom's mass must change – indeed, must decrease – when the atom emits the photons. A change in the atom's energy is accompanied by a change in the atom's mass – and in the same sense, both changes here being a decrease. After having emitted the photons, the atom has less inertia than before.

11.3 Conversions and parallel changes

When mentioning the topic of this chapter, newspapers and popular science often talk about converting this into that or vice versa. Let us see what is actually possible.

Convert matter into radiation

In the 1930s, physics discovered that nature provides us not only with the humble electron but also with its anti-particle, a particle with the same mass as the electron but a positive charge. The electron's anti-particle is called a *positron*. Indeed, most particles in physics come paired with anti-particles. A large "atom smasher" produced the first human-made anti-protons at Berkeley, California, in 1955. The anti-proton has a negative charge but is otherwise quite like a proton. There is also an anti-neutron; its difference from the garden-variety neutron is subtle, for there is no charge whose sign one can merely reverse. Thus, in principle, there is even an anti-atom: positrons would swarm around a negatively charged atomic nucleus consisting of anti-protons and anti-neutrons.

Collections of particles, like electrons and protons, we think of as matter, and the anti-particles count as such, too (though one could also call them "anti-matter").

For decades now, physicists have studied the annihilation that occurs when an electron and a positron get together. The electron and positron disappear, and two or three gamma rays emerge. If we use an arrow to mean "become," then we can write the reaction succinctly as

electron+positron→two or three gamma rays.

The reaction converts matter into radiation.

In principle, one could place in contact an atom and its anti-atom; they would annihilate each other, and many gamma rays would emerge. Thus the reaction

$$\text{atom+anti-atom}\rightarrow\text{lots of radiation}$$

is theoretically possible. This reaction, too, converts matter into radiation.

Convert radiation into matter

Can either of the preceding reactions go the other way? Yes. When a gamma ray enters a block of lead shielding, sometimes the gamma ray produces an electron and a positron and disappears in the process. The reaction is written like this:

$$\text{gamma ray+lead nucleus}\rightarrow\text{electron+positron+lead nucleus.}$$

The lead nucleus plays the role of a catalyst. Thus one can indeed convert radiation into matter. Figure 11.4 shows such a conversion, but the context needs some explanation, as follows.

When an electrically charged particle moves through a liquid, it pulls electrons from neutral molecules (if it is positively charged) and pushes electrons off molecules (if it is negatively charged). Thus a moving charged particle leaves behind an invisible trail of detached electrons and molecular ions. If the liquid were under high pressure and if then the pressure were suddenly released, bubbles would form in the liquid. (Such bubble formation is similar to what happens when you open a bottle of soda or beer.) The detached electrons and the ions provide "seed" locations for bubble formation. Thus the invisible trail suddenly becomes a visible trail of tiny bubbles. The tracks that you see in figure 11.4 were produced in precisely this fashion.

The track pair with a V-shaped apex was formed by an electron and a positron. An intense magnetic field bent the electron's trajectory into a spiral in one direction and bent the positron's trajectory into a spiral in the opposite direction. The gamma ray that entered the liquid and then produced the electron–positron pair has no electric charge, and so the gamma ray left no track as it passed one liquid molecule after another. Although the electron and positron seem to arise out of nothing, they actually arise from the invisible gamma ray. The reaction is an analog of the reaction described in the text above. A nucleus in some molecule of the liquid takes over the role played by the lead nucleus.

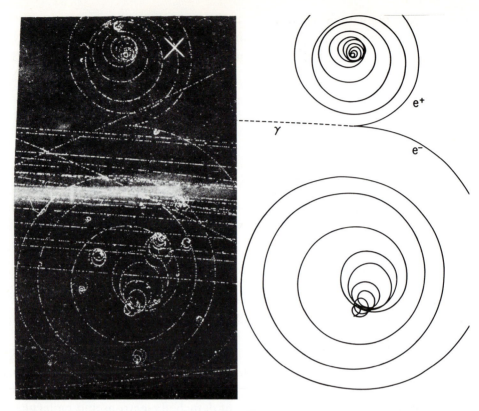

Figure 11.4 Converting radiation into matter. The track pair with a V-shaped apex shows the trajectories of an electron and a positron. The pair of particles was produced by a gamma ray (which left no trail of bubbles in the liquid). The many other tracks are extraneous; most of them were produced by electrically charged pions passing through the liquid. (Photo courtesy of Brookhaven National Laboratory and Clifford Swartz.)

Mass and energy: parallel changes

We have just discussed two kinds of conversions: matter to radiation and radiation to matter. We must distinguish these conversions from the implications of

$$\Delta E = (\Delta m) c^2.$$

Our derivation of this equation in section 11.2 followed a route that Einstein suggested, and our derivation is similar to the route that Einstein himself used, in 1905, when he first derived the connection between ΔE and Δm. In his paper, Einstein noted that when the energy of an atom changes by ΔE, the mass changes by $\Delta E/c^2$ in the same sense. (To see this, divide the equation above on both sides by c^2 and then read it from right to left.) The energy and the mass both decrease or both increase. We have *parallel changes*. There is *no* conversion of one into the other.

244

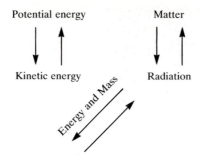

Figure 11.5 Conversions versus parallel changes. The relations in the top line are true conversions. But with energy and mass, we have parallel changes: the equation $\Delta E = (\Delta m)c^2$ tells us that when the energy of a physical system decreases, so does the mass. No conversion occurs.

The notion of parallel changes may be easier to accept if you remember that mass means inertia and is an attribute. For an atom or any object, the attributes energy and inertia change in parallel.

Figure 11.5 summarizes the conclusions of this section.

11.4 Mass and speed

This section investigates another aspect of mass: how mass depends on speed. As always, we have to remember that "mass" means inertia and is an attribute, not a thing.

When we ask how mass depends on speed, we must distinguish two inherently different situations:

(a) the object could be at rest in our frame of reference;

(b) the "object" is never at rest in any (physically realizable) frame of reference.

An electron and a tennis ball belong to situation (a), whereas a photon belongs to situation (b). In this section, we address situation (a) exclusively.

Next, the definition of two symbols:

m_0=mass when object has zero speed;

m=mass when object has speed v as observed in our frame of reference.

When the object has zero speed, it is at rest (as observed by us or by someone accompanying the object), and so m_0 is called the object's *rest mass*. It is an intrinsic property of the object. For example, every electron has a rest mass of 9.11×10^{-31} kilograms. A typical tennis ball has $m_0 = 0.06$ kilograms. (The unadorned letter m serves also as the generic symbol for mass – when one does not specify whether the object is in motion or at rest.)

If we set into motion a tennis ball that is initially at rest, the ball acquires kinetic energy, the energy associated with motion. Its energy increases: $\Delta E > 0$. Our central equation,

$$\Delta E = (\Delta m) c^2,$$

implies that Δm will be positive, too. We deduce that the ball's inertia increases. Thus the mass m of the moving ball is greater than the rest mass m_0. The inequality

$$m > m_0$$

holds for a ball – or any object – with kinetic energy.

For our purposes in this chapter, we don't need to know in detail how m depends mathematically on the speed v. The inequality $m > m_0$ suffices. The exact mathematical form is worked out in appendix B, "How mass depends on speed," and will be needed only in chapter 13.

11.5 From $\Delta E = (\Delta m)c^2$ to $E = mc^2$

Now we have everything we need to make the final step to $E = mc^2$.

Let me write the mass m so that we see symbolically the amount that arises from kinetic energy and the amount that is intrinsic to the object, be it a tennis ball or an electron:

$$m = (m - m_0) + m_0.$$

The difference $m - m_0$ is the extra inertia, the increase in mass that accompanies the kinetic energy. The term m_0 is the intrinsic inertia.

Is there energy associated with the intrinsic inertia, the rest mass?

We could have the object – at rest – annihilate with an "anti-object," producing radiation. Then we could use the radiation to make steam, run a steam engine, and ultimately lift a weight. So, yes, there is energy associated with the rest mass.

Moreover, we could turn the process around, creating the object (along with its anti-object) out of radiation. (For a tennis ball, we would need to create electrons, protons, and neutrons, together with their anti-particles. This is possible, though admittedly not practical.) Every bit of energy that

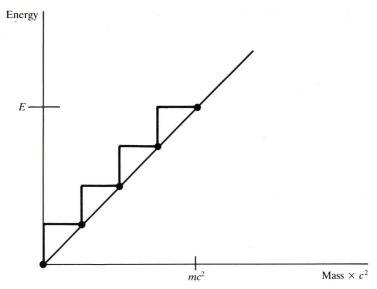

Figure 11.6 We start with zero energy and zero inertia. Every increment ΔE in energy (brought in from the outside) is accompanied by an increment Δm in inertia; the relationship is $\Delta E=(\Delta m)c^2$. A sequence of incremental steps takes the system to its final energy E and final mass m. These final values are proportional to each other, with proportionality constant c^2.

goes into creating the object would be accompanied by an increment in inertia, according to the relation

$$\Delta E = (\Delta m)c^2.$$

Figure 11.6 sketches this symbolically. Starting from zero for *both* the object's energy *and* its inertia and adding up the increments, we emerge with

$$E = mc^2.$$

Einstein saw the equation $E=mc^2$ as the most significant single result of the special theory of relativity. Energy and inertia had been distinct notions – and to some extent they still are – but here is a synthesis. The attribute energy is always accompanied by the attribute inertia; the proportionality constant is c^2. We abuse the language only slightly if we say that inertia is a property of energy.

In 1905, Einstein summarized his theoretical discovery with the sentence, "The mass [that is, the inertia] of a body is a measure of its energy content." Energy (in all forms) has the property of inertia, of reluctance to

247

undergo a change in velocity. The more energy that went into forming a body, the more inertia the body has.

11.6 Nuclear fission

Despite what the newspapers may have led you to believe, the equation $E=mc^2$ is *not* essential for understanding nuclear fission or nuclear weapons. Chemical explosives (like dynamite or TNT) are based on changes in electrical potential energy within atoms and molecules. Nuclear explosives and nuclear reactors are based on changes in electrical and nuclear potential energy within atomic nuclei. The parallel between chemical and nuclear explosives is very close, and in neither case is $E=mc^2$ essential. After all, dynamite was invented by Alfred Nobel before Einstein was born.

Nonetheless, the equation $E=mc^2$ is not totally irrelevant either, and so let us see what goes on in nuclear fission.

The nucleus of a uranium atom may split apart spontaneously, or the splitting may be induced by the absorption of a neutron that happens to strike the nucleus. The latter process is much more common in explosions and reactors; we focus on it.

Figure 11.7 shows what may happen when a uranium nucleus absorbs an extra neutron. A verbal description of induced nuclear fission runs like this:

$$\text{neutron} + \text{uranium} \rightarrow \text{fast-moving fragments} + \text{a few fast} + \text{radiation.} \tag{1}$$
$$\text{nucleus} \quad \text{(like barium and} \quad \text{neutrons}$$
$$\text{krypton nuclei)}$$

After the uranium nucleus has absorbed the neutron, it splits into two large fragments – that is the "fission" – and also spits out some individual neutrons and emits some radiation, typically photons of so high an energy that they fall in the gamma ray range of the electromagnetic spectrum. The large fragments can be classified according to the number of protons each has. In the figure, the splitting yields fragments that are identical to a barium nucleus (with 56 protons) and a krypton nucleus (with 36 protons). Other combinations sometimes occur, such as two palladium nuclei (with 46 protons each) or a xenon nucleus (with 54 protons) and a strontium nucleus (with 38 protons). Rarely are the fragments of exactly the same size; rather, one fragment is typically about 40 per cent more massive than the other. The number of free neutrons varies also, from none whatsoever to as many as half-a-dozen. Typically, two or three neutrons emerge.

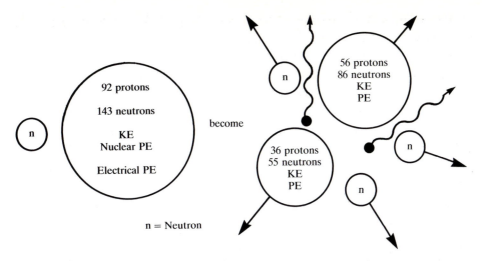

Figure 11.7 Induced fission visualized. Abbreviations: KE=kinetic energy; PE=potential energy. The uranium nucleus is shown as having 143 neutrons and, of course, 92 protons, giving a sum equal to 235. Thus the nucleus is the isotope uranium-235. This isotope undergoes fission readily, whereas the more common isotope, uranium-238, does not. When uranium is mined, the isotope uranium-235 constitutes only 0.7 per cent of the uranium in the ore. Each dot-and-wiggly arrow represents a gamma ray.

The many protons within the original uranium nucleus repel one another and tend to disrupt the nucleus. Only the attractive nuclear forces manage to hold the nucleus together. The absorption of the extra neutron provides the nucleus with extra energy, so that the electrical repulsion of the protons can more easily overcome the attractive nuclear forces and can push opposite halves of the nucleus apart. Once the splitting has occurred, the electrical repulsion between the two large fragments imparts tremendous kinetic energy to the fragments as they accelerate away from each other.

Energy is conserved in the fission process:

$$E_{\text{initial}} = E_{\text{final}},$$

where "initial" and "final" refer to the state of affairs before and after fission. There is no change in the number of protons or neutrons (or electrons, too, if one bothers to include them); rather, they are rearranged into new aggregations. The same protons and neutrons are "packaged" differently, first in one large aggregation, then in two smaller aggregations plus a few free neutrons. Different packaging leads to different amounts of potential energy being stored in the aggregations. Some electrical potential energy is converted to kinetic energy and the energy of radiation (and to nuclear potential energy). The kinetic energy and radiant energy cause the explosion or are used to run a steam turbine and thereby generate electricity.

How do we fit $E=mc^2$ into this scene?

Suppose that, as nuclear engineers, we would like to calculate – in advance – how much energy will become available as kinetic energy and radiation. We start the calculation by invoking conservation of energy:

$$E_{initial} = E_{final}.$$

Then we write things out in more detail:

$$\left(\begin{array}{c} \text{Energy of} \\ \text{incoming neutron} \end{array}\right) + \left(\begin{array}{c} \text{energy of} \\ \text{uranium nucleus} \end{array}\right) =$$

$$\left(\begin{array}{c} \text{energy of fast-moving} \\ \text{fission fragments} \end{array}\right) + \left(\begin{array}{c} \text{energy of} \\ \text{fast neutrons} \end{array}\right) + \left(\begin{array}{c} \text{energy of} \\ \text{radiation} \end{array}\right). \quad (2)$$

Next, we note that the energy of a moving object is a sum: the energy that the object has when at rest plus the object's kinetic energy. The object's rest mass m_0 tells us how much energy the object has when at rest:

$$\left(\begin{array}{c} \text{Energy of object} \\ \text{when at rest} \end{array}\right) = m_0 c^2. \quad (3)$$

This statement is just $E=mc^2$ when specialized to an object at rest. Putting things together, we may write

$$\left(\begin{array}{c} \text{Energy of} \\ \text{moving object} \end{array}\right) = \left(\begin{array}{c} \text{energy of object} \\ \text{when at rest} \end{array}\right) + \left(\begin{array}{c} \text{kinetic} \\ \text{energy} \end{array}\right)$$

$$= m_0 c^2 + \left(\begin{array}{c} \text{kinetic} \\ \text{energy} \end{array}\right). \quad (4)$$

Now we apply equation (4) to reexpress various terms in equation (2). The uranium nucleus was at rest, and so its energy is just its rest mass times c^2. In a nuclear reactor, the neutron that induces the fission is usually a slow neutron, and so we may – for simplicity – ignore its kinetic energy. Hence the neutron's contribution to the initial total energy is just its rest mass times c^2 In short, on the left-hand side of equation (2) we have just the rest mass of the uranium nucleus and that of the neutron, both times c^2.

The situation with the fission fragments is more complex. The fragments and the free neutrons move rapidly, and so both have significant kinetic energy. The radiation contributes energy also. The various contributions to equation (2) lead to the equation

$$\left(\begin{array}{c}\text{Rest mass of}\\\text{neutron}\end{array}\right)c^2 + \left(\begin{array}{c}\text{rest mass of}\\\text{uranium nucleus}\end{array}\right)c^2 = \left(\begin{array}{c}\text{rest masses of fast-moving}\\\text{fission fragments}\end{array}\right)c^2$$

$$+ \left(\begin{array}{c}\text{rest masses of}\\\text{fast neutrons}\end{array}\right)c^2 + \left(\begin{array}{c}\text{all the kinetic}\\\text{energy}\end{array}\right) + \left(\begin{array}{c}\text{energy of}\\\text{radiation}\end{array}\right).$$

To tidy things up, we combine the rest mass terms on each side:

$$\left(\begin{array}{c}\text{Rest masses of neutron}\\\text{and uranium nucleus}\end{array}\right)c^2 =$$

$$\left(\begin{array}{c}\text{rest masses of fission}\\\text{fragments and free neutrons}\end{array}\right)c^2 + \left(\begin{array}{c}\text{all the kinetic}\\\text{energy}\end{array}\right) + \left(\begin{array}{c}\text{energy of}\\\text{radiation}\end{array}\right).$$

Next, we subtract from both sides of the equation the sum of the final rest masses (times c^2):

$$\left(\begin{array}{c}\text{Rest masses of neutron}\\\text{and uranium nucleus}\end{array}\right)c^2 - \left(\begin{array}{c}\text{rest masses of fission}\\\text{fragments and free neutrons}\end{array}\right)c^2 =$$

$$\left(\begin{array}{c}\text{all the kinetic}\\\text{energy}\end{array}\right) + \left(\begin{array}{c}\text{energy of}\\\text{radiation}\end{array}\right). \quad (5)$$

The rest masses of various nuclei and of a neutron have been known from independent measurements for a long time; so we can look up those masses in tables and thereby evaluate the left-hand side. The sum of the energies on the right-hand side is what we set out to calculate. A numerical answer to our question is no farther away than the reference volumes on a library shelf.

Recall that the basic process in nuclear fission is the conversion of electrical potential energy to kinetic energy. How, then, do we verbally interpret the equation above? We know that there is energy associated with rest mass. If the sum of the final rest masses is substantially less than the sum of the initial rest masses, then – after fission – much less energy is associated with rest masses than before. The difference must show up as kinetic energy and the energy of radiation.

What about conversions?

The number of protons and neutrons (and electrons) has not changed. So there is *no* conversion of matter to radiation.

Rather, the rest masses of the aggregations are smaller than before because less electrical potential energy is present in them. It may seem less mysterious if we say, "Because the final aggregations have less potential energy in them, they have less inertia; since 'mass' is just a technical name for inertia, the rest masses of the aggregations are smaller." The true conversion is from electrical potential energy to kinetic energy and the energy of radiation (and to nuclear potential energy). The equation $E=mc^2$ does help us to calculate the amount of energy so converted.

11.7 Processes in nuclear physics

In section 10.3, we talked about radioactive decay and the mean life; fission held center stage in the last section. You might like to get a coherent, broad view of the processes in nuclear physics, and to that end this section is dedicated. Nothing essential to topics farther along in the book emerges in this section; enjoy it for its intrinsic interest.

We can group the processes of nuclear physics into three categories, some of which have sub-divisions.

1 Photon emission and absorption
The atomic nucleus can emit a photon, just as an atom can emit a photon. This nuclear process is the emission of electromagnetic radiation, occurring primarily because the positively charged protons are in motion within the nucleus. The energy of such photons is very high, and they can easily penetrate wood and aluminum. Five centimeters of lead will stop most of them.

When radioactivity was discovered in the last years of the nineteenth century, physicists classified the unknown "radiation" into three types, named alpha, beta, and gamma in order of increasing ability to penetrate material. The gamma "radiation" turned out to be the photons we are discussing, and so these photons are called "gamma rays."

A nucleus can also absorb a gamma ray, just as an atom can absorb a photon of visible light.

2 Rearrangement without change in the number of neutrons or protons
The neutrons and protons in the system under study may be redistributed without any change in the number of each type. Such rearrangement processes may themselves be sub-divided.

Fission, induced or spontaneous. In fission, the nucleus splits into two large fragments and, typically, emits several free, individual neutrons. In the last section, we studied induced fission: a neutron is first absorbed by the atomic nucleus. The absorption provides the nucleus with extra energy, so that the electrical repulsion of the protons can more easily overcome the intrinsically attractive nuclear force and can push opposite halves of the nucleus apart. The process of fission can also occur spontaneously. When the isotope uranium-238 decays, it does so by spontaneous fission in 5 out of every 10 million instances. In the other instances – the vast majority – the isotope decays by the next process in our list.

Alpha decay. In alpha decay, the nucleus emits a tight cluster of two protons and two neutrons. That cluster was originally called an *alpha particle* and is the least penetrating of the three original kinds of "radiation" from radioactive material. A few sheets of paper will stop alpha particles. Today, the alpha particle is recognized as identical to the nucleus of the most common isotope of helium, helium-4. An atom of helium-4 has two electrons swarming around its nucleus and has two protons and two neutrons forming that nucleus.

Absorbing projectiles. To study atomic nuclei, physicists often bombard them with protons, neutrons, alpha particles, or other aggregations of protons and neutrons. Sometimes those projectiles bounce off the target nucleus; at other times they are absorbed. By observing just what happens and how often it happens, physicists learn about the internal structure of the target nucleus and about the force that holds protons and neutrons together.

Under the rubric of "absorbing projectiles" we can include the absorption of the neutron that leads to induced nuclear fission. And uranium is not the only element that absorbs neutrons. To adjust the fission rate in a nuclear reactor, the operator controls the number of slow neutrons present (and ready to be absorbed by uranium and hence induce fission). The isotopes boron-10 and cadmium-113 have a tremendous propensity for absorbing neutrons. Inserting those isotopes (as constituents of control rods) into the reactor core sops up neutrons and reduces the number available for inducing fission. Partially withdrawing the control rods decreases the competition by boron and cadmium for neutrons and gives a higher fission rate.

Fusion. In nuclear fusion, two nuclei are brought together to form a single nucleus. In some ways, fusion is the reverse of fission, the splitting of one nucleus into two others. The electrical repulsion between positively charged nuclei resists fusion; the attractive nuclear force encourages it.

3 Transformation of neutron into proton and vice versa

The last of the nuclear processes is the most bizarre. New particles are formed where none of their type existed before. The two basic processes are these:

$$\text{neutron}\rightarrow\text{proton}+\text{electron}+\text{anti-neutrino,} \qquad (1)$$

$$\text{proton}\rightarrow\text{neutron}+\text{positron}+\text{neutrino.} \qquad (2)$$

The neutrino has no electric charge (as its name suggests) and is believed to travel at the speed of light (always). It has an anti-particle, subtly different from itself. Because the neutrino and anti-neutrino have no electric charge, they are extremely difficult to stop. For the same reason, they do no harm in passing through lead or a human being. The two reactions cited above conserve electric charge. In the first, there is always zero net amount of charge; in the second, always an amount of charge numerically equal to the proton's positive charge.

The beta "radiation" of the turn of the century was found to be emission of electrons, generated by the first reaction. Thus *beta rays* are nothing but garden-variety electrons. For beta rays of the energy typical in nuclear processes, an aluminum sheet a few millimeters thick will stop them handily.

Carbon-14 provides an example of radioactive decay by emission of a beta ray, a process called *beta decay*. A carbon-14 nucleus, with 6 protons and 8 neutrons, turns into a nitrogen-14 nucleus, having 7 protons and 7 neutrons, and the electron and anti-neutrino depart swiftly from the scene. At its leisure, the nitrogen-14 nucleus adds an electron to the six that it inherited from the carbon-14 atom, thus making a neutral nitrogen-14 atom. That extra electron may be stolen from some other atom or may be the electron emitted by some other carbon-14 nucleus in the environment or . . . . The possibilities are legion, but the seventh electron is not likely to be the electron that the nucleus itself emitted.

The sequel to fission provides other examples of decay in which an electron is emitted. Induced nuclear fission produces two large fragments, and they are usually radioactive. For example, the smaller fragment in figure 11.7 is equivalent to a nucleus of krypton-91. That isotope is overly rich in neutrons and starts a decay chain that transforms neutrons into protons, together with the emission of electrons and anti-neutrinos. The sequence goes first to rubidium-91 and thence through strontium-91 and yttrium-91 to zirconium-91, a stable isotope. The larger fragment is equivalent to a nucleus of barium-142. That nucleus is also overly rich in neutrons and decays to lanthanum-142 and then to cerium-142, an isotope whose mean life is so long (7×10^{15} years) that, for all practical purposes, the isotope is stable.

The radioactivity of certain barium isotopes, such as barium-142, was the central clue in the discovery of nuclear fission. A good place to start the story is in Rome in the 1930s, where Enrico Fermi bombarded uranium with neutrons, hoping to make new elements beyond uranium in the periodic table: trans-uranic elements. The idea was that uranium would absorb a neutron, would find itself with too many neutrons to be stable, would transform a neutron into a proton (and emit an electron and an anti-neutrino), and thus would become a new element with 93 protons. Fermi's results were inconclusive: he found plenty of new radioactivity but could not identify the elements from which it came.

In Berlin, Otto Hahn and his colleagues, Lise Meitner and Fritz Strass-mann, set out to repeat some of Fermi's experiments and to sort out the species produced. As chemists – or, as they called themselves, as "nuclear chemists" – they had long experience in the chemical separation and identification of radioactive elements.

The trail of tests led them to a provisional conclusion: when uranium is bombarded with neutrons, new isotopes of radium are formed, perhaps through a sequence of decay steps. This was surprising but not unreasonable, for radium is not far from uranium in the periodic table. Radium is similar to barium in its chemical properties; they precipitate out in the same chemical reactions. Hahn and Strassmann had only a few thousand radioactive atoms to work with at a time, and so when they wanted to extract any "radium" from the original uranium sample, they first mixed in stable barium and then precipitated out a mixture of the radioactive "radium" and the stable barium. When they tried to separate the radioactive "radium" from the barium, it would not separate. After the most stringent of tests, they concluded that the new radioactive "radium" was actually radioactive barium. And this was astonishing, for barium lies more nearly in the middle of the periodic table. With great diffidence, Hahn and Strass-mann suggested that the uranium nucleus was being split into large fragments. Within weeks, Lise Meitner and Otto Robert Frisch wrote the first theoretical paper on the subject and coined the term "nuclear fission." The events are so momentous that the dates are worth noting: Hahn and Strass-mann came to their conclusion in December 1938; Meitner and Frisch sent their paper off in January 1939. In short, it was the winter of 1938–9, less than a year before a troubled Europe would descend into World War II. But back, now, to the nuclear processes themselves.

If you look back to figure 10.7, giving the decay chain for uranium-238, you will see a sequence of alpha decay steps (diagonally leftward) interrupted occasionally by pairs of beta decay steps (diagonally rightward).

Relations (1) and (2) of this sub-section are the basic neutron-to-proton and proton-to-neutron transformations, but variations on these processes occur also. For example, when anti-neutrinos were first detected in the laboratory, the sequence of events was the following. A nuclear reactor produced a prodigious flow of anti-neutrinos by the first of the basic reactions. The anti-neutrinos passed through the reactor walls as though nothing were there. Outside the reactor, a few anti-neutrinos from the flow (about 25 per hour) were absorbed by protons in a cylinder of special organic liquid. (The volume of fluid would have filled a bathtub easily.) In each instance, the proton was transformed into a neutron, and a positron was emitted:

$$\text{anti-neutrino} + \text{proton} \rightarrow \text{neutron} + \text{positron}. \tag{3}$$

The physicists detected the neutron and the positron and thus, indirectly, confirmed the existence of the elusive neutrino and anti-neutrino.

The detection methods provide us with further examples of nuclear processes. The positron annihilated with one of the many atomic electrons already present in the liquid. Gamma rays from the annihilation event were detected at the periphery of the cylinder and provided evidence for the positron. The neutron was absorbed by a cadmium nucleus, present in solution just for that purpose. After the cadmium nucleus absorbed the neutron, it got rid of surplus energy by emitting a gamma ray and then settled down to a new equilibrium. That gamma ray has a characteristic energy (different from the energy of the gamma rays produced by positron–electron annihilation); detection of that gamma ray was evidence for the neutron in relation (3).

11.8 Scales of energy release

Section 11.6 emphasized that the equation $E=mc^2$ is not essential for understanding nuclear fission and its release of energy. Why, then, is that equation so intimately associated in our minds with nuclear fission? To answer the question, we need to examine the amount of energy released in a typical reaction, first chemical and then nuclear.

The burning of carbon in air to form carbon dioxide is a typical chemical reaction. When the carbon atom and the molecule of diatomic oxygen combine to form carbon dioxide, the nuclei do not change, nor does the total number of electrons. Rather, there are changes in electrical potential energy and in the kinetic energy of the electrons and nuclei. These changes enable a candle flame to emit light, that is, to emit energy in the form of light.

When a nucleus of uranium-235 absorbs a slow neutron and undergoes fission, there is no change in the number of protons, neutrons, or electrons. Rather, there is a decrease in the electrical potential energy. The decrease releases energy that appears as the energy of gamma radiation and as the kinetic energy of the two large fragments and the free neutrons. (Simultaneously, the nuclear potential energy increases somewhat; more will be said about that shortly.)

So far, the chemical reaction and the nuclear process appear to be quite similar. Indeed, they are, but now comes a crucial difference: the energy released in one fission reaction is about 100 million times as large as the energy released when one molecule of carbon dioxide is formed. The ratio is 10^8 to 1. That is a stupendous ratio.

Indeed, let me help you to visualize such a ratio. Stretch out your arms sideways and note the distance from one hand to the other. Now, how far away must you look to see a distance about 10^8 times as long as that? Across town? No. Across the state? No. Across the country? No. To the moon? Yes. The distance from you to the moon is about 10^8 times as large as your armspan.

In section 11.6, we used conservation of energy to work out the equation

$$\left(\begin{array}{c}\text{Rest masses of neutron} \\ \text{and uranium nucleus}\end{array}\right)c^2 - \left(\begin{array}{c}\text{rest masses of fission} \\ \text{fragments and free neutrons}\end{array}\right)c^2 =$$

$$\left(\begin{array}{c}\text{all the kinetic} \\ \text{energy}\end{array}\right) + \left(\begin{array}{c}\text{energy of} \\ \text{radiation}\end{array}\right)$$

for nuclear fission. Tables tell us what the difference in rest masses itself is:

$$(3.91981 \times 10^{-25}) - (3.91670 \times 10^{-25}) = 3.11 \times 10^{-28} \text{ kilograms}$$

for the specific fission reaction of figure 11.7. Although small by human standards, this amount is readily measurable in the lab – as the existence of the tables testifies.

The parallel between nuclear fission and burning carbon implies that a similar equation must hold for a chemical reaction. Specifically, we must be able to write

$$\left(m_{0\text{ carbon}} + m_{0\text{ diatomic oxygen}}\right)c^2 - \left(m_{0\text{ carbon dioxide}}\right)c^2 =$$

$$\left(\begin{array}{c}\text{all the kinetic} \\ \text{energy}\end{array}\right) + \left(\begin{array}{c}\text{energy of} \\ \text{radiation}\end{array}\right).$$

Before the reaction, the molecules move so slowly that we may ignore their kinetic energy. Thus the term "all the kinetic energy" correctly describes the kinetic energy after the reaction. In the chemical reaction, the energy on the right-hand side is only the fraction $1/10^8$ of the corresponding energy in nuclear fission. Therefore the difference in rest masses (on the left-hand side) will be smaller by the factor $1/10^8$, and so small a difference in molecular masses *cannot* yet be measured in the laboratory. In short, the equation $E=mc^2$ is not employed in ordinary chemistry because the changes in the rest masses of atoms and molecules are still too small to measure. Although a true equation in chemistry, $E=mc^2$ is not useful there, and so it did not enter public consciousness through chemistry.

How the disparity arises

A single fission reaction, we noted, releases about 10^8 times as much energy as a single chemical reaction. We can readily build a qualitative understanding of why such a huge disparity exists.

The key ingredient is the great difference between the size of an atom and the size of an atomic nucleus. Ever since Ernest Rutherford discovered the atomic nucleus in 1911, physicists have known that the radius of an atom is about 10^5 times larger than the radius of an atomic nucleus. Here is a comparison to dramatize the factor 10^5. If a nucleus were the size of the pencil point in a mechanical pencil (or the ball in a ball-point pen), then the outermost electrons of the atom would be about one football field away.

The electrical force between two electric charges decreases strongly when the separation of the charges increases. Consequently, the electrical force between an outer electron and a proton (in the nucleus) is much, much smaller than the electrical force between two protons in the tiny nucleus.

Chemical reactions, such as the formation of molecules from atoms, affect only the few outermost electrons of the atoms. Such an outer electron is attracted by the protons in the nucleus but is repelled by the inner electrons. The two forces tend to cancel each other. It is as though the outer electron interacted with only one or a few protons. Coming on top of the relatively great separation of outer electron and nucleus, the partial cancellation of forces reduces further the net electrical force on an outer electron.

Now let us look at a typical nuclear fission reaction, the one portrayed in figure 11.7. All 36 protons in the krypton fragment push on all 56 protons in the barium fragment. The total repulsive force between the fragments is $36 \times 56 = 2016$ times as strong as the repulsion between merely two protons (at the same separation).

Thus we have found two reasons why electrical interactions can release much more energy in nuclear fission than in chemical reactions. In the fission process, the charged particles are much closer together (at the start, anyway). Moreover, the constituent protons of each fragment act cooperatively to increase the repulsive force by a factor of about 2000 (relative to what just one proton in each fragment would produce).

What role do specifically nuclear forces play in the fission drama? Primarily, the attractive nuclear forces keep the original uranium-235 nucleus from flying apart immediately (because of the mutually repulsive forces among the protons). Similarly, the nuclear forces restrain the barium and krypton nuclei from catastrophic disintegration. Indeed, in the fission reaction portrayed in figure 11.7, the nuclear potential energy actually increases as the reaction proceeds. The two fission fragments have more surface area than did the original uranium nucleus. Moving protons and neutrons from deep inside a nucleus to the surface layer increases their nuclear potential energy (just as moving a ball higher in the earth's attractive gravitational field increases the ball's potential energy). Consequently, the increase in surface area is accompanied by an increase in nuclear potential energy. Moreover, some neutrons become free neutrons; their escape from the attractive nuclear forces is accompanied by an increase in nuclear potential energy. The electrical potential energy, however, decreases during the fission process, and it decreases more than the nuclear potential energy increases. Thus electrical repulsion is literally the essential reason why nuclear fission can "release energy," converting potential energy into the energy of radiation and into the kinetic energy of the fission fragments and the fast-moving neutrons.

More history

Already in 1905, Einstein recognized that nuclear physics might provide a context for testing and using the equation $E=mc^2$. Toward the end of his paper, he wrote, "It is not impossible that a test of the theory will succeed with bodies whose energy content is variable to a high degree, for example, with radium salts." Einstein had in mind the radioactive decay of elements like radium, thorium, and uranium, then being studied in compounds such as radium chloride, a "salt," just as sodium chloride is our common table "salt." By the 1930s, differences in nuclear rest masses could be measured accurately enough to confirm Einstein's relationship. Moreover, Hans Bethe and others could use the differences to explain the generation of radiant energy in the sun as a nuclear process, primarily fusion. In physics, the equation $E=mc^2$ was well established by 1935.

For the general public, however, $E=mc^2$ did not enter their lives until

1945, when two nuclear bombs were dropped on Japan. Newspapers featured the equation $E=mc^2$, but typically for incorrect reasons. A single nuclear reaction releases about 10^8 times as much energy as a single chemical reaction. That is the primary practical distinction. From it flows the consequence that $E=mc^2$ is useful in calculating some properties of nuclear reactions, while the equation is not yet useful for chemical reactions.

But, of course, there is much more than a non-essential equation to the history of nuclear weapons. In the next section, we look at the origins.

11.9 Leo Szilard

Leo Szilard invented the nuclear chain reaction and drafted the letter that Einstein sent to President Roosevelt, the letter that started America on the road to the atomic bomb. Szilard helped to build the bomb and then spent the rest of his life working to prevent its use. How did all of this come about? This section offers a sketch of the history.

Szilard was born in Budapest, Hungary, in 1898. Though fascinated by physics, he studied engineering there and then went to Germany for advanced work. The physicists in Berlin – an array that included Planck and Einstein – captured his interest so completely that he switched to physics. Szilard taught and practised physics in Berlin until 1933. He got to know Einstein well, and they even took out a patent together on a refrigerator that used liquid metal as a component and pumped it around by electromagnetic means. When Adolf Hitler came to power, Szilard left Germany and moved to Britain.

For Szilard, the sequence of events that led to the atomic bomb started with a book he happened to read. In his words,

> in 1932 while I was still in Berlin, I read a book by H. G. Wells. It was called *The World Set Free*. This book was written in 1913, one year before the World War, and in it H. G. Wells describes the discovery of artificial radioactivity and puts it in the year of 1933, the year in which it actually occurred. He then proceeds to describe the liberation of atomic energy on a large scale for industrial purposes, the development of atomic bombs, and a world war which was apparently fought by allies of England, France, and perhaps including America, against Germany and Austria, the powers located in the central part of Europe. He places this war in the year 1956, and in this war the major cities of the world are all destroyed by atomic bombs. Up to this point the book is exceedingly vivid and realistic. From then on the book gets to be a little,

shall I say, utopian. With the world in shambles, a conference is called in Brissago in Italy, in which a world government is set up.

This book made a very great impression on me, but I didn't regard it as anything *but* fiction. It didn't start me thinking whether or not such things could in fact happen. I had not been working in nuclear physics up to that time.

(References for the quotations in this section appear at the chapter's end, in "Additional resources.")

The seed lay dormant in Szilard's mind for a year. Indeed, as he recalled, it lay there

until I found myself in London about the time of the British Association meeting in September 1933. I read in the newspapers a speech by Lord Rutherford [a Nobel laureate for his early work on radioactivity], who was quoted as saying that he who talks about the liberation of atomic energy on an industrial scale is talking moonshine. This set me pondering as I was walking the streets of London, and I remember that I stopped for a red light at the intersection of Southampton Row. As the light changed to green and I crossed the street, it suddenly occurred to me that if we could find an element which is split by neutrons and which would emit *two* neutrons when it absorbed *one* neutron, such an element, if assembled in sufficiently large mass, could sustain a nuclear chain reaction. I didn't see at the moment just how one would go about finding such an element, or what experiments would be needed, but the idea never left me.

The neutron had been discovered by James Chadwick in 1932. Szilard's idea was that, if two neutrons come out, we can afford to lose one and still have one left to continue the sequence of splittings. And if the loss is less than one on the average, then the splittings will increase in number with each generation.

Szilard knew that he had a powerful idea – and a dangerous one. The steps he took were novel:

In the spring of 1934 I had applied for a patent which described the laws governing such a chain reaction. It was the first time, I think, that the concept of critical mass was developed and that a chain reaction was seriously discussed. Knowing what this would mean – and I knew it because I had read H. G. Wells – I did not want this patent to become public. The only way to keep it from becoming public was to assign it to the government. So I assigned this patent to the British Admiralty.

Szilard pursued his idea, looking for an element that had the desired nuclear properties. All the searches failed, and in December 1938, Szilard wrote to the British Admiralty, suggesting that the patent application be withdrawn because he could not make the process work. A month later, in January 1939, came the news that Otto Hahn and Fritz Strassmann in Berlin had discovered the fission of uranium. That is, when uranium absorbs a neutron, it splits apart. The questions leaped to Szilard's mind: will more neutrons come out? And is a chain reaction possible with uranium?

By now Szilard was in America, permanently. He sent a telegram to the Admiralty, canceling his withdrawal and reinstating the patent application. And he set to work at Columbia University to answer his questions. By the summer of 1939, Szilard and Enrico Fermi had found that neutrons are indeed emitted during the fission of uranium. Whether a chain reaction was possible remained uncertain.

Szilard, already a refugee from Nazi Germany, grew increasingly worried about Europe and the prospect of war. Austria and Czechoslovakia had been swallowed up. What would come next? And what if Germany were to build a nuclear bomb?

The first thoughts of Szilard and Eugene Wigner, a fellow Hungarian physicist, were to warn the Belgian government against selling uranium to Germany. (In those days, Belgium controlled the land that was then called the Belgian Congo and is now the independent country of Zaire.) They went to see Einstein, who was vacationing on Long Island, New York, because Einstein was a friend of the Queen of the Belgians and might be persuaded to write the Queen. When they described to Einstein the possibility of a nuclear chain reaction, his response was, "I never thought of that," but he was quick to see the implications. In the course of the next two weeks, the initial proposal evolved into a decision to warn President Roosevelt about the possibility and dangers of a nuclear chain reaction. As Szilard recalled years later,

> [Edward] Teller and I went to see Einstein, and on this occasion we discussed with Einstein the possibility that he might write a letter to the President. Einstein was perfectly willing to do this. We discussed what should be in this letter, and I said I would draft it. Subsequently, I sent Einstein two drafts to choose from, a longer one and a shorter one.
>
> We did not know just how many words we could expect the President to read. How many words does the fission of uranium rate? So I sent Einstein a short version and the longer version; Einstein thought the longer one was better, and that was the

version which he signed. The letter was dated August 2, 1939. I handed it to Dr. Sachs [an economic advisor to the President, whom Szilard had consulted about political strategy and who had suggested to Szilard that Einstein write Roosevelt] for delivery to the White House.

To dramatize the context in which the letter was written, let me remind you that August 2, 1939, was only a month before September 1, 1939, the date Germany invaded Poland and World War II started. Here are excerpts from the fateful letter:

> F. D. Roosevelt
> President of the United States
> White House
> Washington, DC
> Sir:
> Some recent work by E. Fermi and L. Szilard, which has been communicated to me in manuscript, leads me to expect that the element uranium may be turned into a new and important source of energy in the immediate future. Certain aspects of the situation which has arisen seem to call for watchfulness and, if necessary, quick action on the part of the Administration. I believe therefore that it is my duty to bring to your attention the following facts and recommendations. . . .
>
> This new phenomenon would also lead to the construction of bombs, and it is conceivable – though much less certain – that extremely powerful bombs of a new type may thus be constructed. . . .
>
> I understand that Germany has actually stopped the sale of uranium from the Czechoslovakian mines which she has taken over. That she should have taken such early action might perhaps be understood on the ground that the son of the German Under-Secretary of State, von Weizsäcker, is attached to the Kaiser-Wilhelm-Institut in Berlin where some of the American work on uranium is now being repeated.
>
> Yours very truly,
> A. Einstein

For half a year, nothing happened – except that a committee was set up. Szilard asked for $2000 (to buy graphite), was promised the money, but did not get it. In Szilard's words,

> it is an incredible fact, in retrospect, that between the end of June 1939 and the spring of 1940 not a single experiment was under way

in the United States which was aimed at exploring the possibilities of a chain reaction in natural uranium.

Stymied by the bureaucracy, Szilard turned again to Einstein, and in March 1940, Einstein wrote a second letter to President Roosevelt, gently prodding the government to take further action. (The two letters to the President constitute Einstein's only participation in the atomic bomb project. He did not engage in the research on fission or in the design of a bomb.) By the spring of 1940, money – $6000 – did come forth, and experiments recommenced at Columbia. Work was transferred from Columbia to the University of Chicago in February 1942 – two months after the Japanese attack on Pearl Harbor. Now the project was being pursued in deadly earnest.

Szilard and Fermi had envisaged a test reactor consisting of uranium embedded in graphite. The graphite would slow down the fast neutrons emitted during fission and make them more amenable to absorption by other uranium nuclei. Design proceeded on that basis, and the reactor was constructed in a squash court under the football stands at the University of Chicago (which had dropped intercollegiate football). On December 2, 1942, the control rods were slowly withdrawn, and the reactor demonstrated that a sustained nuclear chain reaction was indeed possible. To the month, it was four years since Szilard had reluctantly concluded that the process could not be made to work. (The basic US patent for the graphite-moderated nuclear reactor cited Szilard and Fermi as the coinventors. The application was filed in 1944, and the patent was issued in 1955.)

The pace picked up as huge plants and laboratories were built: the gaseous diffusion plants at Oak Ridge, Tennessee (to separate the fissile isotope uranium-235 from the much more common isotope uranium-238); the plutonium production reactors at Hanford, Washington; and the bomb design laboratory at Los Alamos, New Mexico. By the end of the war, Szilard's $6000 had grown to an expenditure of two billion dollars.

Now we enter a new phase in this history. As Szilard recalled,

> in the spring of '45 it was clear that the war against Germany would soon end, and so I began to ask myself, "What is the purpose of continuing the development of the bomb, and how would the bomb be used if the war with Japan has not ended by the time we have the first bomb?"

> Initially we were strongly motivated to produce the bomb because we feared the Germans would get ahead of us, and the only way to prevent them from dropping bombs on us was to have bombs in

readiness ourselves. But now, with the war won, it was not clear
what we were working for.

There were new questions, new worries.

Szilard came to oppose using the bomb against Japanese cities. He tried
to reach President Roosevelt with that message, but before he could get
through to the President with another letter – this time via Mrs. Roosevelt –
the President died. Then Szilard tried to reach President Truman and was
sent to see James Byrnes, who had been designated to be Truman's secretary of state. Besides discussing the use of the bomb, Szilard expressed his
concern lest an arms race between Russia and America develop. He
returned to Chicago depressed.

In the aftermath of Szilard's visit to Byrnes, a committee of Chicago
scientists was formed to examine the question of whether the bomb should
be used and, if so, how. The committee was chaired by James Franck, a
Nobel laureate for his work in atomic physics; Szilard was a member. The
committee's report, in Szilard's words,

> advised against the outright military use of atomic bombs in the
> war against Japan. It took a stand in favor of demonstrating the
> power of the atomic bomb in a manner which will avoid mass
> slaughter but yet convince the Japanese of the destructive power
> of the bomb.

This was Szilard's position during the spring and summer of 1945.

Once again events move swiftly. The first nuclear weapon was tested at
Alamogordo, New Mexico, on July 16, 1945. The first bomb was dropped
on Hiroshima on August 6th; the second, on Nagasaki three days later. And
Japan capitulated on August 15th.

The war was over, but – for Szilard – the battles for an enlightened
control of nuclear weapons had only begun. The first of the post-war battles
was over the bill, placed before Congress in the fall of 1945, for the control
of atomic energy, the May–Johnson bill. Szilard went to Washington to
lobby against the proposed legislation. He found Washington keenly interested in the issue and found himself and his colleague, Edward Condon,
greatly in demand as authorities. In Szilard's words,

> We set ourselves a schedule: everybody wanted to see us, and we
> decided that we would keep Cabinet members waiting one day,
> Senators for two days, and Congressmen for three days before
> we'd give them an appointment.

Such was the tactical ability of the opposition (which included others), and the bill never reached the House floor.

Already in 1933 Szilard had seriously proposed to go into biology. His invention of the nuclear chain reaction had changed his mind; physics was fascinating again. But after the war Szilard did transform himself into a molecular biologist. And he kept on lobbying for the control of nuclear weapons. In 1962, Szilard founded the Council for a Livable World, an organization that provides financial support to congressional candidates with enlightened views on the arms race. It is part of his political legacy.

Szilard died in 1964. He had been seriously ill four years earlier and had dictated several whimsical political stories from his hospital bed. The collection was published as *The Voice of the Dolphins*, the title of the first story. In lieu of a preface, Szilard inserted a poem by Stephen Vincent Benét. The poem was written in 1938, before the discovery of nuclear fission. Yet the poem speaks uncannily to us in our nuclear world, speaks as only a poem can. I recommend to you Benét's *Nightmare for Future Reference*.

11.10 Retrospection

It is time to look back over the chapter and collect the basic results.

From two frames of reference, we studied an atom and its emission of photons; we deduced the equation

$$\Delta E = (\Delta m)c^2.$$

Literally, when the atom's energy decreased, so did its mass, its inertia. Einstein explicitly – and we implicitly – generalized to the statement, when any system's energy changes by ΔE, its inertia changes by an amount numerically equal to $\Delta E/c^2$.

Physics recognizes several types of conversions. One may convert potential energy to kinetic energy and vice versa. One may convert matter into radiation (as when an electron and a positron annihilate), and one may convert radiation into matter (as when a gamma ray in lead forms an electron–positron pair). With mass and energy, the situation is radically different: changes in mass and energy occur in parallel, as Einstein pointed out in his first paper. The equation

$$\Delta E = (\Delta m)c^2$$

implies that either *both* the energy and the mass increase or *both* decrease. There is no conversion of one into the other.

When an object at rest – be it electron or tennis ball – is set into motion, its inertia increases. The object has been given extra energy (in the form of kinetic energy), and so the equation $\Delta E = (\Delta m)c^2$ implies that the object's inertia has increased. We agreed to let m_0 denote the inertia that the object has when at rest; thus m_0 is called the object's rest mass. The letter m represents the inertia in general. Now that the object is moving (with speed v, say), our reasoning gives us the inequality

$$\left(\begin{array}{c} \text{Inertia when object} \\ \text{is in motion} \end{array} \right) > \left(\begin{array}{c} \text{inertia when object} \\ \text{is at rest} \end{array} \right),$$

that is, $m > m_0$.

There is energy associated with the rest mass m_0. For example, we can make that energy manifest by having the object annihilate with its anti-object, thus producing radiation. In turn, we use the radiation to make steam, run a steam engine, and ultimately lift a weight. It may be that we can get the energy into an industrially useful form only by destroying the object, but energy is there, nonetheless.

Moreover, we could turn such a process around, creating the object (along with its anti-object) out of radiation. Every bit of energy that goes into creating the object would be accompanied by an increment in inertia, according to the relation

$$\Delta E = (\Delta m)c^2.$$

Starting from zero for both the object's energy and its inertia and adding up the increments, we emerge with

$$E = mc^2.$$

Reading this equation from right to left, we learn that the inertia of an object is proportional to the object's energy.

Consider an atom of common oxygen, the isotope oxygen-16, at rest. In what ways does the atom have energy? Eight electrons swarm around the nucleus. They contribute kinetic energy and the energy associated with their rest masses. There is electrical potential energy arising from the electrons' mutual repulsion and from their attraction to the positively charged nucleus. The nucleus itself has eight protons and eight neutrons, engaged in their own motions in the atom's core. So those sixteen particles contribute kinetic energy and the energy associated with their rest masses. The protons repel one another electrically; so there is more electrical potential energy. Finally, nuclear forces hold the protons and neutrons together, and so there is nuclear potential energy.

All these forms of energy contribute to the inertia of the oxygen atom. Each does so in the same proportion, following the rule $\Delta E = (\Delta m)c^2$.

Inertia is a property of energy – of all forms of energy – and so the more energy that went into forming a body, the more inertia the body has. This sentence captures the essence of the equation $E = mc^2$.

With this insight in hand, we can understand why an atom's mass decreases when the atom emits two photons, as in section 11.2. The energy for the photons comes from the electrical potential energy of the electrons. When that potential energy decreases, it makes a smaller contribution to the atom's inertia, and so the atom – taken as a whole – has less mass.

For practical use, the equation that links energy, rest mass, and kinetic energy is especially valuable. Even an object at rest has energy; the object's inertia when at rest is evidence for that energy. The general equation $E = mc^2$, specialized to an object at rest, implies

$$\begin{pmatrix} \text{Energy of object} \\ \text{when at rest} \end{pmatrix} = m_0 c^2.$$

The energy of a moving object is the sum of the object's energy when at rest plus its kinetic energy. Thus we may write

$$\begin{pmatrix} \text{Energy of} \\ \text{moving object} \end{pmatrix} = \begin{pmatrix} \text{energy of object} \\ \text{when at rest} \end{pmatrix} + \begin{pmatrix} \text{kinetic} \\ \text{energy} \end{pmatrix}$$

$$= m_0 c^2 + \begin{pmatrix} \text{kinetic} \\ \text{energy} \end{pmatrix}.$$

Holding misconceptions at bay

The central equations of this chapter, $\Delta E = (\Delta m)c^2$ and $E = mc^2$, have given rise to wrong interpretations and serious misconceptions. How can you defend yourself against them? The next paragraph offers a shield.

The context is specifically the physics of $E = mc^2$. In this context, remember that "mass" means inertia, a reluctance to undergo a change in velocity, and that inertia is an attribute, not a thing. This provides you with a test: if you cannot sensibly substitute the word "inertia" for the word "mass" in an author's sentence, then the author is misusing the word "mass," and the entire sentence is deeply suspect.

The test will usually distinguish sense from nonsense. On certain subtle issues, however, it will not give a ringingly clear response. A case in point is the phrase "the equivalence of mass and energy." Even Einstein used this phrase (already in 1907). Let us try rewriting it: the equivalence of inertia and energy. Hmmm. What meaning can the word "equivalence" have here? The dictionary offers several possibilities. One is "alike in signifi-cance." Can inertia (a reluctance to undergo a change in velocity) and energy (the ability to do work) be "alike in significance"? Surely not.

The dictionary also offers "equal in value." This meaning is plausible here, for the equation $E=mc^2$ says that mass and energy are proportional to each other, the proportionality constant being c^2. Some historical digging suggests that this is what Einstein meant and what we should mean today. In 1907, Einstein published his third paper on the topic of mass and energy; it carried the title, "On the inertia of energy, as implied by the principle of relativity." In a footnote he remarked casually that a certain equation expresses "the principle of the equivalence of mass and energy," as though such a principle were obvious or well known. Yet this seems to be the first time the phrase appears in the scientific literature. The context suggests that Einstein meant the phrase to denote the universal numerical proportion-ality of inertia and energy. The concepts of inertia and energy remain distinct; their numerical values (for any given system) are proportional, with proportionality constant c^2.

Additional comments about the conceptual issues of this chapter appear in appendix C, "More about $E=mc^2$."

Additional resources

Einstein's first paper on $E=mc^2$ carried the title, "Does the inertia of a body depend on its energy content?" The paper appeared in *Annalen der Physik*, volume 18, pages 639–41 (1905). A translation is available in the anthology *The Principle of Relativity*, consisting of papers by A. Einstein, H. A. Lorentz, H. Minkowski, and H. Weyl and translated by W. Perrett and G. B. Jeffrey, Dover Publications, New York.

A marvelous source of information about Einstein is emerging, volume by volume: *The Collected Papers of Albert Einstein*, edited by John Stachel (Princeton University Press, Princeton, NJ, 1987 and subsequently). Volume 1, *The Early Years*, covers the period 1879–1902; volume 2, *The Swiss Years*, is devoted to the interval 1902–7 and contains Einstein's first paper on $E=mc^2$, together with editorial comment. Translation volumes accompany the primary volumes, in which documents are printed in their original language.

In an article entitled "Did Einstein really discover $E=mc^2$?" W. L. Fadner provides a fine history of the connection between inertia and energy: *American Journal of Physics*, volume 56, pages 114–22 (1988). Moreover, the article provides literature citations for all of Einstein's early papers on the subject. A classic on the same topic is Max Jammer's *Concepts of Mass in Classical and Modern Physics*

(Harvard University Press, Cambridge, MA, 1961). Not all authorities agree, however, with everything Professor Jammer asserts.

All quotations in section 11.9, entitled "Leo Szilard," except the Einstein letter, come from Szilard's "Reminiscences," printed in *The Intellectual Migration, Europe and America, 1930–1960*, edited by Donald Fleming and Bernard Bailyn (Harvard University Press, Cambridge, MA, 1969). The individual quotations come from pages 99, 100–1, 101–2, 113, 117, 122–3, 129 n, and 138, respectively. An expanded version of the "Reminiscences," together with 122 related documents, was published in 1978 as *Leo Szilard: His Version of the Facts, Selected Recollections and Correspondence*, edited by Spencer R. Weart and Gertrud Weiss Szilard (MIT Press, Cambridge, MA, 1978) and available as an inexpensive paperback. The Einstein letter appears among the documents in this collection (on pages 94–6) and also in *The Collected Works of Leo Szilard, Scientific Papers*, edited by Bernard T. Feld and Gertrud Weiss Szilard (MIT Press, Cambridge, MA, 1972). Szilard's little book of political tales was published as *The Voice of the Dolphins and Other Stories* (Simon and Schuster, New York, 1961). Benét's poem, *Nightmare for Future Reference*, forms the introduction to that volume and may also be found in *Selected Works of Stephen Vincent Benét*, volume 1, poetry (Farrar and Rinehart, New York, 1942).

Eugene Rabinowitch painted an incisive portrait of Szilard in his obituary of Szilard and James Franck, both of whom died in May 1964: *Bulletin of the Atomic Scientists*, October 1964, pages 16–20. The breadth of Szilard's interests and his intellectual virtuosity are engagingly recounted by Alice Kimball Smith in "The elusive Dr. Szilard," *Harper's*, July 1960, pages 77–81. Eugene Wigner and Szilard knew each other already as students in Berlin; so Wigner writes about his colleague from long and intimate acquaintance in *Biographical Memoirs, National Academy of Sciences*, volume 40, pages 337–47 (Columbia University Press, New York, 1969). Edward Shils, a social scientist, knew Szilard in the post-war years; they shared an active concern with international politics and arms control. His memoir of the "calmly desperate genius" appeared in *Encounter*, December 1964, pages 35–41. A brief account of what it was like to be a young physicist working with Szilard is given by Bernard T. Feld in *Physics Today*, July 1975, pages 24–5.

The official history of the United States' project to build the bomb is presented by Richard G. Hewlett and Oscar E. Anderson, Jr., in *The New World, 1939/1946*, Volume 1 of *A History of the United States Atomic Energy Commission* (Pennsylvania State University Press, University Park, PA, 1962). Richard Rhodes provides a gripping account of that same project in *The Making of the Atomic Bomb* (Simon and Schuster, New York, 1986).

Einstein on Peace, edited by Otto Nathan and Heinz Norden and published by Simon and Schuster (New York 1960), presents Einstein's long attention to issues of war and peace.

Questions

1. With a pair of identical lasers, we can irradiate a molecule of carbon dioxide from opposite sides simultaneously. Figure 11.8 illustrates this. Suppose the molecule absorbs one photon from each beam.

Figure 11.8 Two laser beams irradiate a molecule of carbon dioxide.

(a) If the laser light has a frequency of $f=1.9\times10^{13}$ oscillations per second, how much energy does each photon have? Recall that Planck's constant h has the numerical value 6.6×10^{-34} joule seconds. A "joule" is the energy unit in the meter-kilogram-second system.

(b) By how much does the energy of the molecule increase? (Two photons, remember?)

(c) By how much does the mass of the molecule increase?

(d) Before absorbing the photons, the molecule had a mass of 7.3×10^{-26} kilograms. By what fraction did the molecule's mass increase? (That is, what is the value of the ratio $\Delta m/m_{initial}$?) Do you think that the change in mass would be easy to measure?

(e) What is the wavelength of the radiation from the laser? Where in the electromagnetic spectrum does such radiation fall?

2. A positron, zipping through some liquid, meets an electron (taken to be at rest and not attached to an atom), and the two annihilate into two gamma rays. Why must the energies of the two gamma rays, when summed together, have a certain minimum value (based on what you have been told)? And what is that value?

(a) First give a verbal and algebraic answer.

(b) Then evaluate your algebraic answer numerically, recalling that the rest mass of an electron is 9.11×10^{-31} kilograms.

3. The isotope hydrogen-2, consisting of a proton and a neutron, has a rest mass of 3.3436×10^{-27} kilograms. The nucleus of ordinary helium, helium-4, has a rest mass of 6.6448×10^{-27} kilograms. Suppose two nuclei of hydrogen-2 fuse to form a single nucleus of helium-4. Would the process, over all, require extra energy from the outside? Or would the process release energy in some form? Explain your reasoning.

4. Here is a list of happenings. For each happening, state whether the rest mass of the object increases, decreases, or remains constant. Explain your reasoning, too.

(a) An atom emits two photons (of equal energy and headed oppositely).

(b) A branding iron is heated to red-hot in a fire.

(c) An archer's bow is stretched.

(d) A flashlight battery runs down while in use.

(e) An atomic nucleus emits two gamma rays (of equal energy and emitted back-to-back).

5. Quantum theory provided us with expressions for a photon's energy and momentum in terms of h, f, and c. If we use the momentum expression to deduce a photon's mass in terms of h, f, and c, is the result consistent with the general relation $E=mc^2$?

6. A proton (initially at rest) can capture a slow neutron and become a hydrogen-2 nucleus. As that nucleus is being formed, a gamma ray is always emitted. The proton and neutron need to get rid of some energy if they are to stay bound together; otherwise they will rapidly come apart again, regaining their original status as a free proton and a free neutron.

(a) Why must the hydrogen-2 nucleus be moving after the gamma ray has been emitted? (In the context of this question, it is appropriate to say that a slow neutron has a negligible speed.)

(b) Consider the entire process described above. Which forms of energy increase? Which forms decrease?

(c) Express the amount of energy that appears as kinetic energy and as the energy of radiation in terms of the rest masses (times c^2) of the particles that are involved.

7. Here is one of our crucial conclusions: when the energy of an atom decreases, the atom's inertia decreases also. We used momentum and momentum conservation to establish that result.

(a) How are the energy and momentum of a photon related? Be precise.

(b) Sketch – both literally and verbally – the line of reasoning we used to establish the "crucial conclusion" cited above. You do *not* need to derive the proportionality constant (that is, c^2). Just get the qualitative result in convincing fashion. No equations are needed or desired. Capture the essentials in words.

8. (*a*) Describe induced nuclear fission. Include a good sketch and write a short paragraph; no equations are needed or desired.

(*b*) What role do electrical forces play in the fission process?

(*c*) Which forms of energy increase during the fission process? And which forms decrease? (By being specific and careful, can you list a total of four?)

9. Here is another "happening": the sun shines. State whether the rest mass of the sun increases, decreases, or remains constant. Also, with an equation and a sentence or two, explain your reasoning.

12   The twins

There seems to be little basis for further arguments about whether clocks will indicate the same time after a round trip, for we find that they do not.

J. C. Hafele and Richard E. Keating,
Science, 1972

12.1 Theory

In this chapter, Alison, who is Alice's twin sister, steps from behind the curtain. While Alison chats with us, Alice prepares for a space voyage. Initially, Alice, Alison, and we are all at rest in a single unaccelerated reference frame. Alison and we remain at rest in that frame, but Alice accelerates up to speed v, travels far away, turns around quickly, travels back, and decelerates to rest once again. Figure 12.1 shows this sequence schematically.

At the instant of Alice's departure, she and Alison are the same age. The question posed to us is this: when Alice returns, will she be as old as Alison has become? Why, of course, because they're twins – but then again, this is a chapter on relativity theory, and we know better than to take anything for granted. Let us investigate the question circumspectly.

Four events are essential to our analysis. Here they are.

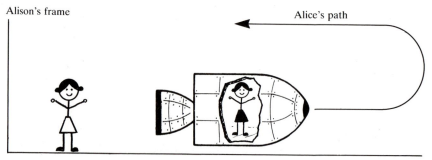

Figure 12.1 Alison (whose skirt is of a pale color) remains at rest in an unaccelerated reference frame while Alice travels far away but then returns.

Event 1. Alice departs – but only after she and Alison have synchronized their watches and agreed that, as twin sisters who have
lived together all their lives, they are of equal age.

Event $2_{traveler}$. Having reached the far point of her journey, Alice
turns her spaceship around and heads home.

Event 2_{home}. Where (in space and time) Alison is when, as we
measure things, Alice turns around.

Event 3. Alice arrives home. She and Alison compare the times that
have elapsed on their watches and compare their ages.

Figure 12.2 displays these four events. The mode of display is new to us:
the horizontal axis represents location in space, but the vertical axis
represents "location in time" as perceived by us. Alison has remained at
rest, and so her location in space is always the same. Her trajectory through
space and time, called her *worldline*, consists of a vertical straight line. The
line commences with event 1, passes through event 2_{home}, and ends (but
only for our purposes!) at event 3.

Alice's trajectory through space and time is quite different. After her
departure (at event 1), Alice moves away from home as time goes on, and
so her worldline is tilted to the right. At event $2_{traveler}$, Alice reverses her

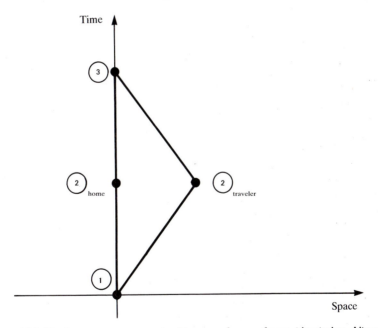

Figure 12.2 The four events, as perceived in our reference frame, identical to Alison's.
Only one space dimension is shown – the direction along which Alice travels out and back
– but times are now shown, along the vertical axis.

direction of travel; as she heads homeward, her worldline is tilted to the left. Two straight but differently tilted segments of worldline constitute Alice's trajectory from event 1 to event 2_{traveler} and thence to event 3.

Our job now is to compare time intervals between various events as perceived by Alice and Alison. We can use our own observations as an intermediary. The central relationship that we need is the time dilation relationship, derived in section 10.1:

$$\Delta t_{\text{any other frame with relative speed } v} = \frac{1}{\sqrt{1-(v^2/c^2)}} \, \Delta t_{\text{frame where events occur at the same place}}. \quad (1)$$

We apply this equation to various event pairs, such as the pair 1 and 2_{traveler}.

For Alice, the events 1, 2_{traveler}, and 3 all happen "at her," and so those events happen at "the same place" for her. After multiplying equation (1) on both sides by the square root and rewriting the subscripts, we have the relation

$$\sqrt{1-(v^2/c^2)} \, \Delta t_{\text{our frame}} = \Delta t_{\text{Alice, the traveling twin}}. \quad (2)$$

The speed v is Alice's speed relative to our (and Alison's) frame.

Because Alison and we remain at rest in the unaccelerated reference frame, Alison and we perceive the same time intervals between the events 1, 2_{home}, and 3:

$$\Delta t_{\text{our frame}} = \Delta t_{\text{Alison, the home twin}}. \quad (3)$$

Table 12.1 provides a tabulation of time intervals. For numerical definiteness, Alice's outward journey is specified to take ten years, as perceived by us. The total elapsed times, we find, are not equal. As equations (2) and (3)

Table 12.1. *Time intervals as perceived by us, Alison, and Alice. The times are cited in years. Equations (2) and (3) enable us to compute the entries in the fourth and third column, respectively, from the entries in the second column.*

Event pair	$\Delta t_{\text{for us}}$	$\Delta t_{\text{home twin}}$	$\Delta t_{\text{traveling twin}}$
1 to 2_{traveler}	10		$\sqrt{1-(v^2/c^2)}\times 10$
1 to 2_{home}	10	10	
2_{traveler} to 3	10		$\sqrt{1-(v^2/c^2)}\times 10$
2_{home} to 3	10	10	
Total elapsed time:		20	$\sqrt{1-(v^2/c^2)}\times 20$
If $v=0.8c$, then total elapsed time:		20	12

imply, Alison and we agree on the total time elapsed between events 1 and 3, but Alice perceives an elapsed time that differs by the factor $\sqrt{1-(v^2/c^2)}$ from what we perceive. Thus the comparison takes the general form

$$\Delta t_{\text{Alice, the traveling twin}} = \sqrt{1-(v^2/c^2)}\ \Delta t_{\text{Alison, the home twin}}. \tag{4}$$

The square root is some number less than 1. Thus, when Alice and Alison compare watches, they find that Alice's watch reads less elapsed time than Alison's does. Alison has become older than Alice. The physiological aging of a human being increases with increasing chronological age. We may suppose that Alice and Alison were well matched at tennis when Alice departed. In a game after Alice's return, we can expect Alison to be less nimble on the court than Alice.

In short, the effect is not just restricted to "clocks," whose behavior we can put at a great psychological distance from us. Rather, the effect directly influences a person's age, in the colloquial meaning of that word, wrinkles and all.

12.2 Experimental test

It would be too much to ask you to believe the different aging of Alice and Alison without my offering some experimental support. Experiments with human twins are not feasible, for reasons that you can readily imagine: to be sure that extraneous effects do not invalidate the experiment, one would need long times (of many years' duration) and high speeds, comparable to the speed of light. Space flights are not up to such demands. We will have to settle for clocks.

Atomic clocks

Physicists have devised remarkably precise and reliable clocks: the atomic clocks. Today, the world's primary time standard is based on atomic clocks whose essential element is a beam of cesium atoms. While in free flight, the cesium atoms can absorb and emit microwaves of a characteristic frequency. Whereas the "second" of time was originally defined in terms of the earth's daily rotation, now the "second" is defined as the time required for 9 192 631 770 consecutive oscillations of the microwave radiation from free cesium atoms. In principle, time keeping has evolved from counting rotations of the earth to counting the oscillations of an electromagnetic wave in the microwave region of the spectrum.

A portable version of the basic cesium beam clock is available. Such a clock is about the size of a suitcase and can be buckled easily into a regular passenger seat on a commercial jet airplane.

The experiment

In October 1971, J. C. Hafele and Richard E. Keating flew a set of four atomic clocks around the world, first eastward and then westward. Figure 12.3 provides a sketch. Before and after each flight, the clocks' readings were compared with the time registered by another set of atomic clocks – the "master clock" – at the United States Naval Observatory, located in Washington, DC.

Here are the *differences* in the measured elapsed time, given as (*elapsed time registered by traveling clock*) – (*elapsed time registered by clock in Washington*):

Traveling westward:

$$\Delta t_{\text{traveling westward}} - \Delta t_{\text{Washington}} = +273 \times 10^{-9} \text{ seconds}$$
$$(\pm 7 \times 10^{-9} \text{ seconds});$$

Traveling eastward:

$$\Delta t_{\text{traveling eastward}} - \Delta t_{\text{Washington}} = -59 \times 10^{-9} \text{ seconds}$$
$$(\pm 10 \times 10^{-9} \text{ seconds}).$$

Regardless of whether we can explain them or not, these are remarkable results. (Note. The four clocks differed a bit among themselves, and this

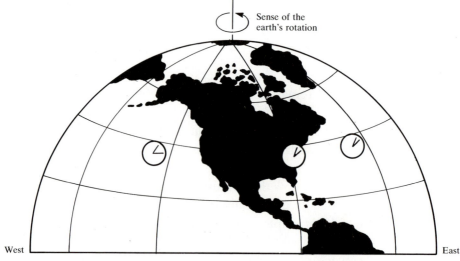

Figure 12.3 A view of the earth from space. Starting from Washington, DC, a commercial jet airplane flew the atomic clocks first eastward around the world and then westward.

accounts for the uncertainties cited with the ± symbol. The uncertainty, you will note, is always small relative to the average value.)

The analysis

Special relativity theory takes as its reference frames only frames that are in uniform motion. Accelerated reference frames are not allowed. A point fixed on the earth, such as Washington, is carried by the earth's rotation first in one direction and then – 12 hours later – in the opposite direction. Such a change in the direction of motion implies accelerated motion, and so Washington does not provide a frame acceptable to special relativity theory. We should view the traveling clocks and the master clock in Washington from the vantage point of outer space.

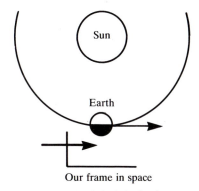

Our frame in space

Figure 12.4 As seen from a nearby star (which looks down on the solar system), our chosen frame moves with the earth's orbital velocity.

Imagine that you look at the solar system from a nearby star. Figure 12.4 illustrates the view. Over the course of the experiment, the earth's orbital velocity – the velocity of the earth's annual motion around the sun – remains constant (at least for practical purposes). We adopt a frame of reference that moves with the earth's orbital velocity. In making this choice, we can be confident that we adopt a frame acceptable to special relativity theory. Moreover, when we view the earth from that frame, the only motion left for the earth is its daily rotation, a desirable simplification.

Two speeds will directly enter our analysis:

v_{earth}=speed with which the earth's daily rotation carries Washington around in space;

v_{jet}=speed of jet airplane relative to the ground.

Table 12.2 displays the speeds of the master clock and of the clocks aboard the jet as they are observed from our frame in space.

Table 12.2. *The speeds of the clocks as seen from our frame in space. Both v_{earth} and v_{jet} are small relative to the speed of light, and so, when we combine them, we may use just the commonsense method: subtract one from the other or add one to the other.*

Clock	Speed as see from space	Relative size of speed
Westward traveling	$v_{earth} - v_{jet}$	Smallest
Washington, DC	v_{earth}	Middle
Eastward traveling	$v_{earth} + v_{jet}$	Largest

Table 12.3. *Our predictions for the lapsed times, stated comparatively.*

Clock	Value of $\sqrt{1-(v^2/c^2)}$	Elapsed time
Westward traveling	Largest	Largest
Washington, DC	Middle	Middle
Eastward traveling	Smallest	Smallest

Now we compare these clocks with a hypothetical clock at rest in our frame in space. That clock plays the role of the "at home" twin. To make predictions, we can use equation (4) of section 12.1 in the form

$$\Delta t_{real\ moving\ clock} = \sqrt{1-(v^2/c^2)}\ \Delta t_{hypothetical\ "at\ home"\ clock}. \tag{1}$$

Each elapsed time that we seek to predict appears on the left in equation (1). It is the time that elapsed between the start of a trip at Washington and the trip's end there. We may schedule the westward and eastward flights to take equal times according to the clock in Washington and hence equal times according to the hypothetical "at home" clock. Then there is only one numerical value for the Δt of the hypothetical "at home" clock. To get predicted readings for the real clocks, we multiply that single value by the square root in equation (1), a root that is different for each real clock. Table 12.2 provides us with the relative sizes of the speed v, the clock's speed. Then table 12.3 lists the elapsed times comparatively, as we predict them. Thus we expect

$$\Delta t_{westward} - \Delta t_{Washington} > 0$$

because we are subtracting the middle value from the largest, and we expect

$$\Delta t_{eastward} - \Delta t_{Washington} < 0$$

because now we are subtracting the middle value from the smallest. These predicted inequalities agree with the experimental differences. This is excellent evidence that our explanation – time dilation in the context of round trips – is on the right track.

To make predictions in full quantitative detail, one must include gravitational effects on the clocks. We shan't go into that supplement to what we have already done. Suffice it to say that the complete professional analysis finds excellent agreement between theory and experiment.

12.3 Reflection

An experiment is not the end of the story. It behooves us to examine two aspects of this peculiar aging.

No longer identical

The first point to make is that Alice and Alison really do have different experiences. That is superficially obvious – Alice gets to ride in a rocket ship, and Alison does not – but it is true also from the narrower perspective of special relativity theory. Alice undergoes acceleration three times: when she sets off on her journey, when she turns around, and when she slows down to land and rejoin Alison. Alice directly perceives these accelerations: the seat in her rocket pushes on her back, or the seat belt restrains her as she turns her rocket around, and the belt does so again as she lands. Alison, remaining at rest in the original reference frame, experiences no pushes or restraints. Alice's acceleration breaks the symmetry that had existed between the identical twins. Of itself, this does not imply that Alice and Alison will have different ages upon Alice's return. Rather, it means we *cannot* argue that Alice and Alison *must* have the same age. Because they now differ in one respect (having undergone acceleration or not), they may differ in other respects as well.

Alice's acceleration is crucial to distinguishing her aging from Alison's, but the basic phenomenon is the time dilation that we learned about in chapter 10. Alice could take twice as long a journey but with the same acceleration; she would just coast for twice as long between the brief acceleration intervals. Then equation (4) of section 12.1 would tell us that Alice and Alison's ages differ by twice as much. The accelerations are crucial for two reasons: (1) they get Alice back so that she and Alison can directly compare their ages, and (2) they end the identicalness of the twin's lives. The accumulated difference in ages, however, depends on the speed v and the length of the journey, not on the acceleration *per se*.

Analogy

Alice and Alison start at the same point in space and time – at event 1 – and rejoin each other at a second point in space and time, event 3. They take different paths through space and time, as already sketched in figure 12.2. When they meet again, they find that they have aged differently: the time intervals that their clocks record are different. This is strange – there is no denying that – but it is analogous to situations that strike us as commonplace.

For example, imagine two hikers. At dawn, they start from their camp at the foot of the mountain. One takes the direct route to the summit, clambering over huge boulders, fun but slow. The other walks half-way around the mountain to a gently-inclined ridge and saunters up it to the summit. The hikers have lunch together at the top. They have ascended the same vertical distance, but the distances walked and scrambled are different. In short, the elevation gained is independent of which route up the mountain is taken, but the distance hiked depends on the path chosen.

So it is with "travel" in space and time. The time interval measured by a traveler – the traveler's personal elapsed time – is a path-dependent quantity. For a fixed pair of events, some journeys between the events will take less time – as measured by the traveler – than other journeys.

Additional resources

J. C. Hafele and Richard E. Keating report their experiment in two articles in *Science*: volume 177, pages 166–8 and 168–70 (1972). A careful presentation of the full theory, including gravitational effects, appears in J. C. Hafele, *American Journal of Physics*, volume 40, pages 81–5 (1972).

A full ten articles on the twins appear in *Special Relativity Theory, Selected Reprints*, published for the American Association of Physics Teachers by the American Institute of Physics, New York, 1963. My favorite paper is Hermann Bondi's "The space traveller's youth."

Questions

1. The first human trip to the moon took about three days (approximately 3×10^5 seconds) each way. The distance from the earth to the moon is roughly 4×10^8 meters. When they returned, how much younger were the astronauts than their twin brothers who remained on earth?

Assume the trip to be made at constant speed (and ignore the gravitational effects that we have not studied). Unless your calculator keeps more decimal places than mine does, you will need the approximation for $\sqrt{1-(v^2/c^2)}$ described in question 5 of chapter 10.

2. An enchanted prince, aged twenty years, has been turned into a frog and can turn back into a prince only if kissed by a beautiful princess on his twenty-first birthday. The only beautiful princess around is fifteen years old and is not allowed to look upon a man until she is twenty-one. How far must the frog-prince travel, on how fast a spaceship, to return to the earth on the day that both he and the princess turn twenty-one? Compare the trip with the distance to our nearest star, Alpha Centauri, which is 4×10^{16} meters (or 4.3 light years) away.

(Note. I find that $\sqrt{1-(v^2/c^2)}$ must equal 1/6. The number of seconds in a year is about 3×10^7, as you can check.)

3. Section 9.5 introduced the idea of synchronizing a set of clocks. Please review that section. Then use the results of this chapter to argue that method number 1 of section 9.5 would indeed retain synchronization (at least to any level of accuracy – short of perfection – that was prescribed). (Note that slow speed is no bar to moving a clock a long distance; one just needs to budget a long time for moving it.)

4. At departure, Alice cannot accelerate up to speed v instantaneously, nor can she turn her spaceship instantaneously, and – finally – she cannot decelerate instantaneously to zero speed to greet Alison. She can, however, do all these things gently and yet in times short relative to the entire trip.

(a) Sketch carefully Alice's worldline in this more realistic description of her trip out and back.

(b) Why is our quantitative conclusion about a difference in elapsed times for Alice and Alison at least an excellent approximation? Why does our qualitative conclusion about aging remain the same?

5. Alison and Alice have digital watches of identical construction. Before Alice's departure, each twin checks her pulse and finds a rate of 74 beats per minute.

(a) Alice checks her pulse again while one-third of the way through her trip, when she is moving at speed $v=0.8c$ relative to Alison. What does Alice find for her pulse rate now? Why?

(b) If Alison could observe Alice measuring her pulse, what rate would Alison find for Alice's pulse rate? Why?

6. Bob and his twin brother Bert purchase new Fords in Los Angeles; each car's odometer registers zero. Bob drives northeast to Denver, Colorado, and thence northwest to Seattle, Washington. Bert takes a leisurely drive northward – up the Pacific coast – and meets Bob in Seattle.

  (*a*) When Bob and Bert compare odometer readings in Seattle, will the two cars still register equal values?

  (*b*) In which ways is the Bob & Bert story analogous to the Alice & Alison saga? In which ways different?

13 The Lorentz transformations

From youth onwards, my entire scientific effort was directed to deepening the foundations of physics.

Albert Einstein,
interview on 6 February 1924

13.1 Deriving the Lorentz transformations

Relativity theory is provocative. It incites an urge to show that the theory is inconsistent or that one *can* beat the speed of light. In this chapter we examine both issues. As a prelude, we need to study how Alice and Bob assign coordinates to an event, that is, how they specify the location and time of an event. Then we need to learn how those assignments are related. Once we have those relations, we can compute how differently Alice and Bob perceive the velocity of some object. And that is a major step toward trying to beat the speed of light.

Alice and Bob specify the location of an event – a firecracker going off, say – by measuring distances from the origins of their respective reference frames. Figure 13.1 displays this. The distance Bob assigns is denoted x; the distance Alice assigns is called x'. Alice specifies the time by using a clock that is present at the event, is at rest in her frame, and is synchronized with the master clock at the origin of her frame. Bob does the same, using a clock at rest in his frame. This procedure – using clocks at the event itself – is a matter of convenience and is something we discussed earlier, in section 9.5. The times are denoted t for Bob's value and t' for Alice's. The two pairs, (x and t) and (x' and t'), are the coordinates of the event for Bob and Alice, respectively.

We do need a further agreement on the setting of clocks before we can relate the coordinates assigned by Alice and Bob. We may specify that the master clocks at the origins both read zero as the origins pass each other. This is just a matter of setting each master clock forward or back to zero. We are free to specify this, and it simplifies matters a bit.

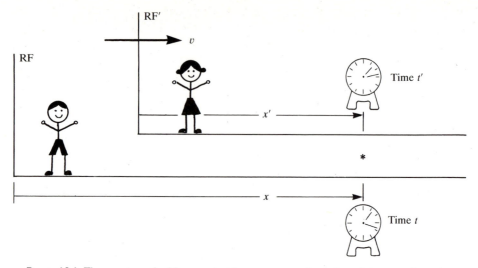

Figure 13.1 The event, marked by an asterisk, occurs away from the origins, out along the direction of relative motion. Bob assigns to the event the coordinates x and t; Alice, x' and t'. Only for the sake of visual clarity is Alice's frame displaced vertically from Bob's.

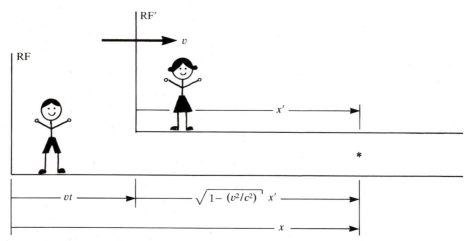

Figure 13.2 Expressing Alice's length coordinate x' in terms of Bob's coordinates x and t. Note that the symbol x' does not represent length on the paper but rather length as measured in frame RF', moving with respect to the printed page.

Alice's length coordinate in terms of Bob's coordinates

Our first task is to see how we may express Alice's length coordinate x' in terms of Bob's length and time coordinates, x and t, for the event. Figure 13.2 shows the essential lengths. We may ask Bob to note where the origin of Alice's frame is at the time – for him – of the event. The origin of Alice's frame has moved at speed v for a time t, and so it is distant by vt from the origin of Bob's frame. The other length that completes the distance to the event is Bob's measurement of what Alice calls length x'. Thus the length x' is analogous to the length of the trout at rest in Alice's frame in section 10.5. As in that section, Bob measures a length contracted by the factor $\sqrt{1-(v^2/c^2)}$. Thus we have

$$x = vt + \left(\begin{array}{c} \text{Bob's measurement of what} \\ \text{Alice calls length } x' \end{array} \right)$$

$$= vt + \sqrt{1-(v^2/c^2)} \ x'.$$

Subtracting vt from both sides and then dividing by the square root gives us

$$x' = \frac{1}{\sqrt{1-(v^2/c^2)}} \ (x - vt). \tag{1}$$

When v/c is much less than 1, the square root is close to unity, and the relationship conforms to our commonsense, aided (perhaps) by a glance back at figure 13.2.

Bob's length coordinate in terms of Alice's coordinates

Next, we take a different perspective and see how we may express Bob's length coordinate x in terms of Alice's length and time coordinates, x' and t'. From Alice's point of view, Bob's reference frame is moving to the left, as displayed in figure 13.3. We may ask Alice to note where the origin of Bob's reference frame is at the time – for her – of the event. Then a glance at the diagram tells us that we may write

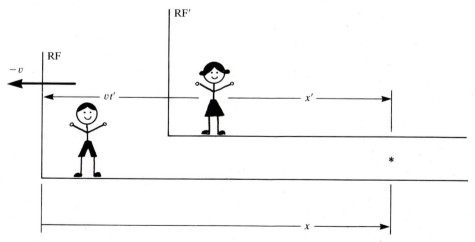

Figure 13.3 Expressing Bob's length coordinate x in terms of Alice's coordinates x' and t'.

$$x' + vt' = \left(\begin{array}{c}\text{Alice's measurement of what}\\\text{Bob calls length } x\end{array}\right)$$

$$= \sqrt{1 - (v^2/c^2)} \ x;$$

the step to the second line follows because Alice measures a contracted value for what Bob calls length x. Division by the square root produces the equation

$$x = \frac{1}{\sqrt{1 - (v^2/c^2)}} \ (x' + vt'). \tag{2}$$

This relation is nicely similar to the relation in equation (1). The sole difference in structure is the + sign in place of the − sign in the numerator, and that change merely reflects the difference in the directions of relative motion.

Alice's time coordinate in terms of Bob's coordinates

To express Alice's time coordinate t' in terms of Bob's coordinates x and t requires no further subtle reasoning on our part. All the information that we need is present in equations (1) and (2). We can use equation (1) to substitute for x' in equation (2) and then solve for t'. The substitution gives

$$x = \frac{1}{\sqrt{1 - (v^2/c^2)}} \left[\frac{1}{\sqrt{1 - (v^2/c^2)}} \ (x - vt) + vt' \right].$$

Collecting all the x and t terms on the left-hand side yields

$$x\left[1 - \frac{1}{1 - (v^2/c^2)}\right] + \frac{1}{1 - (v^2/c^2)} \ vt = \frac{1}{\sqrt{1 - (v^2/c^2)}} \ vt'.$$

The coefficients of x can be put over a common denominator; the numerator is simply $-v^2/c^2$. Then division of both sides by the coefficient of t' produces the result

$$t' = \frac{1}{\sqrt{1 - (v^2/c^2)}} \left(t - \frac{vx}{c^2}\right). \tag{3}$$

The term in vx/c^2 is novel and probably unexpected. But perhaps we should have expected something bizarre, for the relations among coordinates must be such that both Alice and Bob observe light to move at the same speed c, and that – though by now familiar – is certainly bizarre. Shortly – and again in section 13.2 – we will see that the term vx/c^2 is precisely what is needed to ensure that the relations reproduce the constancy of the speed of light.

Right now we can note that the vx/c^2 term expresses the relativity of

Table 13.1. *The Lorentz transformations. The relative motion of the two reference frames is along their x and x' axes. The primed frame RF' moves to the right with speed v relative to frame RF.*

$$x' = \frac{1}{\sqrt{1-(v^2/c^2)}}(x-vt) \qquad\qquad x = \frac{1}{\sqrt{1-(v^2/c^2)}}(x'-vt')$$

$$t' = \frac{1}{\sqrt{1-(v^2/c^2)}}\left(t-\frac{vx}{c^2}\right) \qquad\qquad t = \frac{1}{\sqrt{1-(v^2/c^2)}}\left(t'+\frac{vx'}{c^2}\right)$$

$$y' = y \qquad\qquad\qquad\qquad\qquad y = y'$$

simultaneity. Suppose *two* events occur, and suppose Bob perceives them to be simultaneous; he assigns the same value of t to each. If those events are spatially-separated, so that Bob assigns different values of x to them, then equation (3) implies that Alice will assign unequal values of her time t' to them. For Alice, the events will not be simultaneous. Note that if the events occur far apart for Bob, then the term vx/c^2 remains significant, even if the ratio v/c is small.

Other relations

To derive an expression for Bob's time coordinate t in terms of Alice's coordinates x' and t', we could again apply algebra to equations (1) and (2), but a quicker route is available. Alice and Bob differ only in the direction of their motion relative to each other: Bob sees Alice moving to the right, while Alice sees Bob moving to the left. So the expression we seek must look like equation (3) but with $-v$ replaced by $+v$. It has been duly entered in table 13.1.

For events that lie off the x-axis, at the level of Alice's head, say, we need to introduce a coordinate perpendicular to the relative motion. Let that be y for Bob and y' for Alice. Because lengths perpendicular to the relative motion are measured to be the same by Alice and Bob, as we derived in section 10.2, those two coordinates are equal: $y'=y$.

Perspective

The relations listed in table 13.1 connect Alice's and Bob's coordinates for an event. With those relations, we can transform Bob's coordinates into Alice's. We just insert Bob's coordinates, such as $x=21$ meters and $t=7\times10^{-8}$ seconds, and also the speed v, such as $v=2\times10^8$ meters/second (which implies that $v/c=\frac{2}{3}$). Then out pop Alice's values for the coordinates of the same event:

$$x' = \frac{1}{\sqrt{1-\left(\dfrac{2}{3}\right)^2}}\;(21 - 2 \times 7) = 9.39 \text{ meters};$$

$$t' = \frac{1}{\sqrt{1-\left(\dfrac{2}{3}\right)^2}}\;\left(7 \times 10^{-8} - \frac{2 \times 21}{3 \times 3} \times 10^{-8}\right) = 3.13 \times 10^{-8} \text{ seconds.}$$

The value of x' is less than x because Alice's frame has been moving rightward since the origins passed each other, and so the event occurs closer to her origin than to Bob's. The value of t' differs from t because ... well, because Alice and Bob must *disagree* on times (as well as on lengths) if they are to *agree* on the speed of light.

Indeed, suppose a burst of light were emitted rightward at "the origin" as the origins passed each other. At time $t=7\times10^{-8}$ seconds, Bob would find the burst to be at a location $ct=(3\times10^8)\times(7\times10^{-8})=21$ meters away from his origin. For Alice, we just calculated that the burst would have traveled 9.39 meters in 3.13×10^{-8} seconds. For her, the ratio of "distance traveled" to "elapsed time" would be

$$\frac{9.39}{3.13 \times 10^{-8}} = 3 \times 10^8 \text{ meters} / \text{second},$$

and so Alice, too, would find the burst to move at speed c.

The second column in table 13.1 enables us to transform Alice's coordinates into Bob's.

The relations in the table were derived by the Dutch physicist Hendrik Antoon Lorentz in a series of papers spanning the years 1892–1904, and for that reason the relations are called the *Lorentz transformations*. Lorentz regarded the relations as just a mathematical tool in his study of electromagnetism. He did not believe the relations represented the actual way Alice's and Bob's coordinates for an event are related. Independently of Lorentz, Einstein derived the relations in his great relativity paper of 1905, using what we have called Principles 1 and 2. Einstein meant the relations to be literally true, referring to measurements by real clocks and meter sticks, just as we do in our derivation.

Let us cast a discerning eye back over our derivation and see what the essential ingredients were. We used length contraction explicitly, and we used the Principle of Relativity, saying that Alice's motion as perceived by Bob and Bob's motion as perceived by Alice differ only in the replacement of v by $-v$; the magnitudes of their relative velocities are the same. Of course, our earlier derivation of length contraction used time dilation. In turn, time dilation used the constancy of the speed of light. We could push the constancy back to Principles 1 and 2, but it suffices to say that the Lorentz transformations follow from the Principle of Relativity (our Principle 1) and the constancy of the speed of light.

13.2 Combining velocities

We know how Alice and Bob relate the coordinates of an event, and so we can figure out how they compare the velocities (for them) of any moving object.

For our purposes, we may take the "object" – be it automobile or electron or burst of light – to move along the direction of Alice and Bob's relative motion. Figure 13.4 sketches this situation. We denote by u the velocity Bob measures for the object; the symbol u' denotes the velocity Alice measures. By using u and u' for these velocities, we avoid confusion with the symbol v that denotes Alice's velocity relative to Bob.

The algebra will be least if we specify that the object passes the origin of Alice's reference frame and the origin of Bob's frame when those origins pass each other. Thus the object passes "the origin" at "time zero" for both Alice and Bob. Each of them can determine the object's velocity by noting the time and distance coordinates of a later event, when the object

Figure 13.4 The "object" – here sketched as an automobile – moves along the direction of Alice and Bob's relative motion.

has traveled some distance from their respective origins. A figure for such an event has already been drawn for us in figure 13.1.

Both Alice and Bob may compute the object's velocity as the ratio (distance traveled)/(elapsed time):

$$u = \frac{x}{t} \quad \text{for Bob;}$$

$$u' = \frac{x'}{t'} \quad \text{for Alice.}$$

To relate the velocity u to the velocity u', we substitute for x and t in terms of x' and t', using the Lorentz transformations listed in table 13.1:

$$u = \frac{x}{t} = \frac{\dfrac{1}{\sqrt{1-(v^2/c^2)}}(x'+vt')}{\dfrac{1}{\sqrt{1-(v^2/c^2)}}\left(t'+\dfrac{vx'}{c^2}\right)}.$$

The factors with the square roots cancel. Next, we factor t' from the denominator and divide that t' into the numerator, remembering that $u'=x'/t'$:

$$u = \frac{u'+v}{1+\dfrac{vu'}{c^2}}. \tag{1}$$

This equation gives us the velocity Bob perceives (u) in terms of the velocity Alice perceives (u') and Alice's velocity relative to Bob (v). For us, equation (1) is the *relativistic rule for combining velocities*.

If either v or u' is very small relative to c (or if both are), then the denominator is approximately unity, and we find that u is approximately the sum of u' and v. This is the commonsense result: the car's velocity relative to Bob is the sum of the car's velocity relative to Alice plus Alice's velocity relative to Bob.

If, however, neither u' nor v is small relative to c, then our result for combining velocities takes on novel features. What does it say if the "object" is a burst of light? If Alice notes the burst to move at speed c, so that $u'=c$, then Bob notes a speed

$$u = \frac{c+v}{1+\dfrac{vc}{c^2}} = \frac{c\left(1+\dfrac{v}{c}\right)}{1+\dfrac{v}{c}} = c.$$

Thus Bob, also, finds the burst to move at speed c. Of course, this is not for us a novel result. It is the constancy of the speed of light again – but its appearance here is a good check on the consistency of the theory. The speed c begets c.

Note that, in equation (1), a divisor different from 1 is crucial. The peculiar term vx/c^2, which first appeared in equation (3) of section 13.1, plays an essential role: it enables the mathematics to incorporate faithfully the constancy of the speed of light.

We can do more with the central result of this section – the rule for combining velocities – and to that we turn in the next section.

13.3 Trying to beat the speed of light

If we start with an object – automobile or electron – that is moving at less than the speed of light, we have two ways in which to try to get the object up to the speed of light – or beyond that speed. We proceed to examine each method in turn.

Run away from it

If the object is moving to the right, say, then one way for us to increase its speed (relative to us) is for us to run to the left. From the point of view of someone watching our antics, the distance between us and the object increases more rapidly than before. We can look back over a shoulder and note how fast the object is now traveling relative to us.

We can let Bob do the running and Alice do the watching, as sketched in figure 13.5. After all, from Alice's point of view, Bob moves to the left (with velocity of magnitude v), and the car moves to the right with velocity u'. When Bob looks over his shoulder, he perceives the car to move at velocity u, given by equation (1) of the preceding section. Let us suppose the car moves with speed $\frac{2}{3}c$ as noted by Alice, so that $u'=\frac{2}{3}c$, and that Bob runs (relative to Alice) at speed $v=\frac{2}{3}c$. Then, for Bob, the car has speed

$$u = \frac{\frac{2}{3}c + \frac{2}{3}c}{1 + \frac{2}{3} \times \frac{2}{3}} = \frac{\frac{4}{3}c}{1 + \frac{4}{9}}$$

$$= \frac{\frac{4}{3}c}{\frac{13}{9}} = \frac{12}{13}c.$$

Figure 13.5 Trying to beat the speed of light by running away from an already-moving object. (Bob is shown running to the left – that is how Alice perceives him – but Bob regards himself as at rest in his own reference frame, frame RF.)

Shucks! The second term in the denominator is big enough to pull down a promising $\frac{4}{3}$ to a mere $\frac{12}{13}$, and that number is less than 1. For Bob, the car still travels at a speed less than c.

You may try other values for u' and v. So long as each value is less than c (which is a rule of the game), then their combination to form the velocity u will yield a speed less than c. (A route to an algebraic proof is offered in question 1 at the chapter's end.)

Succinctly, no matter how we combine velocities that are less than c in magnitude, we cannot achieve a speed equal to c or greater than c.

Push on it

The other method for trying to beat c is straight-forward: we push on the object, making it go faster. A continuous push will produce a continuous increase in speed. Will this method do the trick?

As the speed increases, so does the object's kinetic energy. The equation $\Delta E = (\Delta m)c^2$ tells us that the object's mass increases concomitantly. Because the object's inertia increases, the same force produces less acceleration. Whether we can push the object up to speed c is cast into question.

To proceed further, we need the detailed dependence of mass on speed. Appendix B, "The dependence of mass on speed," derives the expression

$$m = \frac{m_0}{\sqrt{1-(v^2/c^2)}}.$$

Here m_0 is the object's rest mass, its inertia when it is at rest. The speed v is the object's speed (as perceived in our reference frame), and m denotes the object's inertia when it is traveling with speed v. If we imagine v to approach the speed of light, then the denominator approaches zero, and the object's inertia grows infinitely large.

The object's momentum, as mass times velocity, is

$$\text{momentum} = \frac{m_0}{\sqrt{1-(v^2/c^2)}} \, v.$$

If we imagine v to approach c, then the momentum grows infinitely large, too.

These preliminaries set the stage. Now we can reason as follows. A force of finite size acting for a finite interval of time can increase the object's momentum by only a finite amount. Thus such a force cannot accelerate the object up to the speed of light (or beyond).

Reasoning about energy leads to the same conclusion. If the speed were to increase to c, not only would the inertia rise to infinity, but also – by the equation $E=mc^2$ – the energy would become infinitely large. In finite time, an object moving at finite speed travels only a finite distance. A finite force acting over a finite distance can impart only a finite amount of energy. Thus the object cannot acquire the infinite amount of energy that travel at speed c would demand.

The growth of inertia with speed is so rapid – as the speed approaches the value c – that no object can be pushed up to speed c. Traveling at speed c is a lonesome distinction. An object may be born traveling at speed c, but it can never be joined by objects created with a lower speed. (To the extent that a photon is an "object," it exemplifies an object "born traveling at speed c.")

Figure 13.6 provides experimental support for the conclusion of this section.

13.4 The chapter in a nutshell

It is time to review; here is the chapter in a nutshell.

The observers Bob and Alice reside in reference frames that move relative to each other. Each observer labels an event according to measurements of distance and time in his or her frame. A connection must exist between the coordinates (x and t) that Bob assigns to the event and the

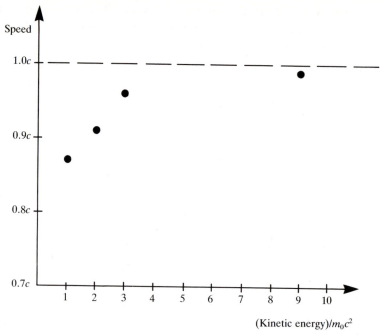

Figure 13.6 The speed of electrons as a function of their kinetic energy. Note that the scale on the vertical axis starts at 0.7c, not at zero speed. Electric fields accelerated the electrons to high speed; thereafter, they were allowed to coast, and their speed was measured. When the kinetic energy equalled $m_0 c^2$, where m_0 is the electron's rest mass, the experiment found the speed to be 0.87c. Despite great increases in kinetic energy – as great as 800 per cent – the speed increased only slightly and approached c as a limiting value. William Bertozzi describes the experiment in the *American Journal of Physics*, volume 32, pages 551–5 (1964).

coordinates (x' and t') that Alice assigns. The Lorentz transformations, displayed in table 13.1, provide the connection.

The most surprising feature of the Lorentz transformations is the presence of the space coordinate (in the form vx/c^2 or vx'/c^2) in the equation relating the time coordinates (t and t') of an event. The constancy of the speed of light generates that presence. In turn, that presence makes simultaneity a relative notion.

Once one knows how Alice and Bob relate their coordinates for an event, one can calculate how they compare the velocities they observe for any object. The relationship, equation (1) of section 13.2, is this:

$$u = \frac{u' + v}{1 + \dfrac{vu'}{c^2}}.$$

The relativistic rule for combining velocities has an unexpected denominator, but the extra term in the denominator ensures that both Alice and Bob measure the speed of a burst of light to be c.

Trying to beat the speed of light is a challenge. Two methods are available: (1) run away from a moving object or (2) push on it. For the first method, the relativistic rule for combining velocities predicts failure. The combination of any two velocities that are less than c in magnitude generates another velocity less than c in magnitude. And if one velocity is already equal to c in magnitude, then the composition just gives c again.

The crucial element in evaluating the second method is the growth of mass – of inertia – with speed. The equation $\Delta E = (\Delta m)c^2$ implies that as the speed and hence the kinetic energy increase, so does the mass. The inertia grows so rapidly as the speed approaches c that no finite force, acting during a finite time and over a finite distance, can push an object up to the speed of light.

The speed of light, we find, is the "ultimate speed." If an object is once observed to move at a speed less than c, then it will always be observed to move at a speed less than c. One may change reference frames (within the bounds of physical realism) and one may push on the object, but the object's perceived speed will remain less than c. If something is ever observed to move at speed c, it will always move at that speed. Obviously, light is one such "something." Almost certainly there are others. Neutrinos have long been believed to move at the speed c, and in that respect they are analogous to photons. Certain gravitational phenomena, called "gravitational waves," are predicted to propagate at speed c.

Thus nature provides us with a unique speed c. Some things, like electrons or spaceships, may strive to attain the speed c; they may come close to their goal, but they are destined never to reach it. Some other things, like electromagnetic radiation and (probably) neutrinos, always move (in vacuum) at the speed c.

Additional resources

Z. G. T. Guiragossián and colleagues compared the speed v of electrons whose energy was $(3 \times 10^4) \times m_0 c^2$ with the speed c of gamma rays. Theory predicts $(c-v)/c = 5 \times 10^{-10}$. The physicists found $|c-v|/c$ to be at least as small as their experimental resolution of 2×10^{-7}. The report appears in *Physical Review Letters*, volume 34, pages 335–8 (1975).

In another report, entitled "Precision experimental verification of special relativity," D. Newman and colleagues provide an annotated list of experimental tests. Their paper appears in *Physical Review Letters*, volume 40, pages 1355–8 (1978).

Questions

1. Here is an algebraic route to the first conclusion in section 13.3. Divide both sides of equation (1) of section 13.2 by c and then subtract the resulting equation from the equation "1=1." On the right-hand side, place all terms over a common denominator. Can you factor so that your equation has the following form?

$$1 - \frac{u}{c} = \frac{\left(1 - \frac{u'}{c}\right)\left(1 - \frac{v}{c}\right)}{1 + \frac{vu'}{c^2}}.$$

What conclusion about the value of u/c can you deduce from this form?

2. Suppose Alice moves relative to Bob at merely the speed of a jet airplane. Then $v/c = 10^{-6}$, a small value. Specify that an event occurred shortly after the origins passed each other but at a very distant location, in the vicinity of Alpha Centauri, the star nearest the sun. In particular, suppose Bob assigns to the event the coordinates

$$t = 150 \text{ seconds},$$

$$x = 4 \times 10^{16} \text{ meters}.$$

Work out the coordinates that Alice assigns to the event. Does the term in vx/c^2 make a significant contribution to the numerical results?

3. For context, please look again at figure 13.4 and review the relativistic rule for combining velocities, equation (1) of section 13.2. Then compute the velocity u in each of the following situations, accompanying each calculation with a sketch that shows the directions of v, u, and u'.

    (*a*) $u' = 500$ meters/second;
        $v = 200$ meters/second.
        (The value 500 meters/second is approximately the speed of air molecules and hence is approximately the speed of sound. Jet planes move roughly this fast.)
    (*b*) $u' = 0.9c$;
        $v = 200$ meters/second.

(c) $u'=0.9c$;
$v=0.9c$.

What verbal conclusions can you draw from these computations?

4. In chapter 8, pions provided a light source moving faster than $0.99c$. We talked about pions because it is impractical to accelerate a flashlight up to such a speed. Now you can show this yourself.

(a) Estimate the rest mass m_0 of a typical flashlight (containing two D-cells). For comparison, a pound of butter has a rest mass of about 0.5 kilograms.

(b) How much kinetic energy would the flashlight have if it moved at speed $v=0.99c$? (Chapter 11 and appendix B contain the relations you need.)

(c) If that energy were to be provided by a large commercial power plant, for how long would all the energy output from the plant have to be poured into accelerating the flashlight? The power plant delivers electrical energy at the rate of 10^9 joules per second. Express your final answer in years.

5. The speed of light in moving water. For light moving through vacuum, Alice and Bob always agree on its speed; it is c, equal to 3×10^8 meters/second. What about light moving through water?

Specify that the water is at rest in Alice's frame and hence moves with speed v as observed by Bob. For Alice, light moves through the quiescent water at speed $u' = \frac{3}{4}c$, a fact we can extract from the index of refraction and the wave theory. Use the relativistic rule for combining velocities to determine the speed of the light (in moving water) as measured by Bob. What conclusions can you draw?

6. A small explosion occurs on the earth. Then, 800 seconds later, earth-bound astronomers see a similar explosion on Mars (which is 1.5×10^{11} meters away on that day). Someone suggests that a flying saucer, on a reconnaissance mission from outer space, set off each explosion as it flew past the earth and then Mars.

(a) Draw a sketch of the context. It will help you to picture the exercise.

(b) How long did it take the light from the explosion on Mars to reach the telescopes on earth?

(c) What was the time interval between the two explosions, as measured in a reference frame in which the earth is at rest (at least so far as its annual revolution about the sun is concerned)?

(d) Could these events have been set off by a single flying saucer?

7. The data in figure 13.6 provide a quantitative test of relativity theory as well as a qualitative confirmation of its predictions.

(a) Work out algebraically the connection between an electron's speed and its kinetic energy. (You can express the total energy E in terms of m_0c^2 and the speed v. If you then subtract from E the energy that the electron possesses when at rest, you will be near your goal.)

(b) Pick any two of the data points in the figure and see whether they support your prediction from relativity theory (when you insert into your algebraic connection either the speed or the kinetic energy and then extract the numerical prediction for the other).

8. Figure 13.7 shows two reference frames in relative motion. Alice's frame moves with speed $v=\frac{3}{5}c$ relative to Bob's frame. When the origin of Alice's frame passes the origin of Bob's frame, clocks at both origins read zero.

In his frame, Bob notes the following coordinates for events 1 and 2.

Event 1: $x_1=5$ meters; $t_1=2\times10^{-8}$ seconds.

Event 2: $x_2=10$ meters; $t_2=3\times10^{-8}$ seconds.

Use the Lorentz transformations (where appropriate) to answer the following questions.

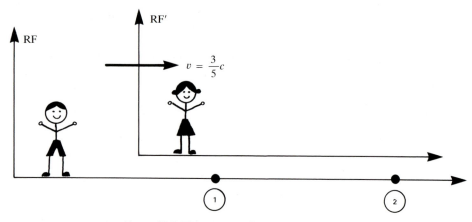

Figure 13.7 The vertical offset is for clarity only.

(a) Do you find that $1/\sqrt{1-(v^2/c^2)}$ equals $\frac{5}{4}$? (I hope so.)

(b) What are the coordinates of event 1 as measured in Alice's frame?

(c) Similarly, what are the coordinates of event 2?

(d) I find $x_2'-x_2'=4$ meters. Do you concur?

(e) Does Alice note the two events to be simultaneous?

(f) If Alice were moving (relative to Bob) 10 per cent faster than specified here, would she perceive the two events to be simultaneous? What if she were moving 10 per cent slower?

9. Two spaceships, A and B, move relative to the earth with speeds of $\frac{4}{5}c$ in opposite directions, as shown in figure 13.8.

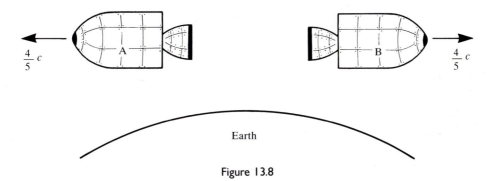

Figure 13.8

(a) According to an observer on the earth, how much does the separation of the spaceships increase in a time of 1 second measured on her clock?

(b) What is the speed of spaceship A according to an observer on B? (You can approach this question as a matter of "combining velocities.")

(c) According to an observer on B, how much does the separation of the spaceships increase in a time of 1 second, as measured on his clock?

(d) In what sense do these questions form a consistency test for relativity theory? Does the theory pass the test?

14 Space and time

Time is that great gift of nature which keeps everything from happening at once.

C. J. Overbeck

Chapter 9 began with the assertion, "Space and time are different from what you thought they were like." By now, probably you agree. This brief chapter draws together verbally the remarkable properties of the space and time in which we live.

The interconnection of space and time

Space and time are intrinsically different from each other. We determine locations in space by laying off a wooden meter stick or by stretching a metal tape, but we determine "locations in time" by reading a clock, be it a pendulum clock or a digital watch or a cesium beam atomic clock. Such an intrinsic difference was appreciated before Einstein's 1905 relativity paper, and it persists to this day. Nonetheless, relativity theory showed space and time to be more closely linked than people had thought. Let us review the connections.

The linkage of space and time shows up most clearly when Alice and Bob – in motion relative to each other – compare the locations in space and time that they assign to the same events. Our investigation of the relativity of simultaneity provides an excellent context. Look again at figures 9.6 and 9.7. Focus first on the event of light's reaching marker (a), fixed in Alice's frame. For Bob, the location in space of that event depends on both the distance of marker (a) from the origin of Alice's frame and on how long she has been moving since the origins of the two frames passed each other. Thus the spatial location of the event for Bob depends on both the spatial location and the time of the event as noted by Alice.

We arranged things so that the events of light reception at markers (a) and (b) would be simultaneous as perceived by Alice. Bob, however, finds that the event at marker (a) occurs before the event at marker (b). Thus, for Bob, the times of those two events depend not only on their times as observed by Alice but also on their spatial locations as noted by her.

In short, the location of an event in both space and time as observed by Bob depends on the location of that event in both space and time as noted by Alice.

Time intervals

Observers in relative motion will usually not agree on the time interval between a given pair of events.

Relativity of simultaneity. We just reviewed the essentials in our derivation of the relativity of simultaneity. Although Alice finds the two reception-of-light events to be simultaneous, Bob does not. Simultaneity is a relative notion – relative to the frame of reference – and so there is no absolute, global "now." A set of events, all of which Alice perceives to occur at one instant of time, "now" for her, would occur for Bob at a variety of times.

Time dilation. To show the relativity of simultaneity, we start with a pair of events that occur at the *same time* but at different spatial locations for Alice and show that those events occur at different times for Bob. Time dilation arises in the opposite context: we start with a pair of events that, for Alice, occur at the *same place* but at different times and show that those events occur farther apart in time for Bob, who is in motion relative to Alice. In this sense, the time interval has become "dilated" for Bob. Succinctly, an observer for whom two events occur at the same place notes the least elapsed time between the events.

Path-dependence of a traveler's personal time. In chapter 12, Alice and Alison, the twin sisters, journey in space and time from one event to another. The first event is Alice's departure in her rocket ship; the other is Alice's return. The time that has elapsed for Alice is less than the time that has elapsed for Alison. Thus the time that elapses for a traveler in space and time depends on the path taken between the starting and concluding events. In short, a traveler's personal time is a path-dependent quantity.

Length contraction

The length of an object, such as Alice's trout, is also a relative notion. Alice, for whom the trout is at rest, notes the longest length. Bob, for whom the trout is moving, measures a shorter length. For him, the trout's length is contracted relative to the length measured in the trout's own frame of reference.

303

Speeds

Speed, as (distance traveled)/(elapsed time), is a notion that grows out of space and time. Peculiarities in the properties of space and time will impose peculiarities on speed or limitations on it.

The constancy of the speed of light. Both Alice and Bob, though moving relative to each other, measure the same speed for the light (traveling in vacuum) from any given source. The value, denoted by the letter c, is always 3×10^8 meters per second.

Speeds less than c persist. We cannot take an object with initial speed less than c and get the object up to the speed of light or beyond that speed. Chapter 13 examined the two basic methods that we can try and found that each fails. If we run away from an already-moving object and observe it over a shoulder, it will indeed be going faster – but no amount of sprinting on our part will get the speed, relative to us, up to c. Nor can we push the object up to the speed c. The growth of inertia with increasing speed is so rapid that every realistic force fails to accelerate the object up to the speed of light.

Success

Every time you listen to the radio or watch television, you receive further confirmation that Maxwell's theory of electromagnetism is correct. In a similar fashion, every day that huge "atom smashers" operate across the world – in Europe, the Soviet Union, the United States, and Japan – they provide further confirmation that Einstein's special theory of relativity is correct. The theory is essential in the design of both the machines and the experiments done with them. For the particles that are accelerated, the relation of speed to energy conforms to the expectations of the theory. The times that elapse between various events exhibit time dilation and path dependence. You can be confident that the properties of space and time collected in this chapter are real and that the "theory" from which they were derived is here to stay.

The continuity of the trail of light

Here, at the end of our excursion, we can look back and review the trail we took. Chapters 1–7 explored the nature of light and culminated in the wave-particle duality. Light is a subtle phenomenon with complementary aspects: wave-like and particle-like. Chapter 8 provided a transition. We asked, does the speed of light depend on the motion of the source of light? Thus we turned to experiments to ask about another property of light, but then we used the answer to develop a theory of space and time. Chapters 9–13 were on the nature of space and time (for the most part), but that nature was discovered by us as a consequence of the constancy of the speed of light. We deduced the constancy by combining Principle 1, the Principle of Relativity, with Principle 2, our result from chapter 8. In short, we built the special theory of relativity – a theory of space and time – on the foundation of experiments about light.

We have learned a great deal about how light behaves and what the properties of space and time are. Absent have been questions like "why does light have a dual nature?" and "why does a burst of light have the same speed for all observers?" Answers to questions about *how* something behaves are often within our grasp – by experiment. Answers to "why" questions are much more elusive. Newton recognized the practical distinction and made it part of his methodology, a methodology that proved extremely fruitful in the ensuing three centuries. Nonetheless, physicists *do* look for the answers to "why" questions. I have not offered you answers to the "why" questions that I just raised (and that may well have occurred to you, also) because compelling answers are not yet known. Surely the trail of light stretches on, and someday one will be able to travel farther. It is something to look forward to.

Glossary

Certain words appear in colloquial speech as well as in physics; this is both a blessing and a curse. You may be familiar with the words but may associate connotations or even denotations that do not belong with the words when used in physics. So take these definitions seriously. They are meant to distinguish usage in physics from usage at the supermarket or in a traffic jam.

ACCELERATION: the rate at which velocity changes with time. Thus motion with constant velocity is *un*accelerated motion. Motion with no acceleration is sometimes called "uniform motion."

AMPLITUDE: the height of a crest above the wave's mid-line.

ATOM: a neutral composite particle formed by electrons swarming around an atomic nucleus.

COMPONENTS: the arrow-like pieces (each with direction and magnitude) which, when added together tail to head, reproduce the originally-given arrow-like quantity (having both a direction and a magnitude).

ELECTRIC FIELD: no brief description is adequate; please see section 5.1.

ELECTROMAGNETIC WAVE: a wave made out of electric and magnetic fields. Light is an electromagnetic wave.

ELECTRON: a tiny negatively charged particle. Electrons form the outer portions of an atom; their motion in a copper wire in your house constitutes an electric current.

ENERGY: the ability to do work, for example, to lift a weight. Energy is an attribute (or "property") of a physical object or of whatever is contained in a specified region of space.

EVENT: an event is anything that happens at some definite location at some definite time (as perceived from some reference frame). Examples: your birth and the first human step on the moon.

FORCE: a push or a pull.

FREQUENCY: the number of oscillations per second (at a fixed location).

HALFLIFE: the time required for one-half of a collection of unstable nuclei or particles to decay.

INTERFERENCE: a fundamental consequence of the superposition principle, whose details are the following. When two crests meet, they reinforce each other, and so a larger crest arises; this is called "constructive interference." When a crest and a trough meet, they cancel each other; this is called "destructive interference."

ISOTOPE: a nucleus (or atom) with specific numbers of both protons and neutrons. For example, a nucleus with 92 protons and 143 neutrons constitutes an isotope of uranium: uranium-235. A nucleus with 92 protons and 146 neutrons constitutes a different isotope of uranium: uranium-238. The word "isotope" comes from the Greek roots "*isos*," meaning "the same," and "*topos*," meaning "place." The various isotopes of uranium appear in "the same place" in the periodic table of elements, namely, the box for all forms of uranium: 92 protons and any number of neutrons.

KINETIC ENERGY: the energy associated with motion.

MAGNETIC FIELD: no brief description is adequate; please see section 5.3.

MASS: inertia; sluggishness; the inherent reluctance to undergo a change in velocity. Mass is an attribute, not a thing. It is an attribute (or "property") of a physical object or of whatever is contained in a specified region of space.

MATTER: tangible stuff; what you can hold in your hand. Note that all matter has mass. Something that has mass, however, need not necessarily be made of matter. Thus the concepts "mass" and "matter" are distinct.

MEAN LIFE: the average lifetime of an unstable isotope or species of particle.

MOMENTUM: the product "mass times velocity."

MUON: an unstable charged particle, about 200 times as massive as an electron.

NEUTRON: a tiny particle similar to a proton but electrically neutral. The mass of a neutron is 0.14 per cent larger than the mass of a proton.

NORMAL: the perpendicular to a surface.

NUCLEUS (ATOMIC): an atomic nucleus consists of a tightly-packed swarm of protons and neutrons, rather like a cluster of honey bees. (The sole exception is the isotope hydrogen-1, which has only one proton and no neutrons.)

PHOTON: a particle of light – but *without* some of the properties we normally associate with the idea of a "particle."

PION: an unstable charged particle, about 270 times as massive as an electron. Pions come in three varieties: positively charged, negatively charged, and neutral. The neutron pion decays into two gamma rays.

PLANE POLARIZED LIGHT: an electromagnetic wave (in the visible portion of the electromagnetic spectrum) whose electric field always oscillates in the same plane.

POSITRON: a particle like an electron but positively charged. The positron is the electron's anti-particle.

POTENTIAL ENERGY: energy that has the "potential" for being converted into kinetic energy.

PROPAGATION DIRECTION: the direction in which a wave moves (or "propagates").

PROTON: a tiny positively charged particle. The nucleus of a hydrogen atom consists of one proton. The electric charges of protons and electrons are equal in magnitude but opposite in sign. The proton is about 2000 times more massive than an electron.

REFERENCE FRAME: a firmly-constructed set of axes for establishing the location in space of an event and one or more clocks for establishing the time of the event.

REFRACTION: the bending of a light beam when it passes obliquely from one substance into another (or passes either into vacuum or out of vacuum).

REST FRAME: a reference frame from which the specified object is perceived to be motionless, that is, at rest. From an outside point of view, a reference frame moving with an object provides a rest frame for that object.

REST MASS: the inertia of an object when it is at rest and hence has zero speed.

SPEED: speed specifies how fast something is moving but ignores the direction of motion. Speed=(distance traveled)/(elapsed time), provided the motion is uniform or the elapsed time is quite short.

SUPERPOSITION PRINCIPLE: to get the resultant pattern when two waves overlap, add the original two patterns algebraically.

UNIFORM MOTION: motion with constant velocity and hence with no acceleration.

VELOCITY: velocity specifies the direction of motion and the speed, that is, it answers the questions "whither?" and "how fast?"

WAVE: a pattern, usually a moving pattern and often a periodic one. Because a wave *per se* is not a thing, there must be something out of which the wave is formed.

WAVELENGTH: the distance between adjacent crests in a periodic wave.

WORK: a measure of effort exerted, specifically, the product of force applied times the distance over which it acts (provided the force is constant and is directed along the distance traveled).

WORLDLINE: an object's trajectory through space and time.

Appendix A Energy

The word *energy* has many colloquial meanings, and it has also a precise technical meaning in physics. This appendix develops the latter meaning as fully as we shall need it in the text.

The route to a technical definition of energy passes through the ideas of force and work before arriving at energy.

Force

In physics, a *force* is a push or a pull. Actually, we have a natural sense for this definition. We push a child's swing or pull a stubborn tent stake out of the ground and know that we are exerting a force.

Work

Imagine yourself lifting a heavy suitcase into an overhead luggage rack, indeed, lifting it two or three times in the course of a trip. You would be doing a lot of work, at least in the colloquial sense. The heavier the suitcase, the more force you would have to exert to lift it. And the higher the luggage rack, the more effort you would have to expend.

In physics, the technical term *work* preserves these commonsense attributes. We focus on the simplest situation, entirely adequate for our purposes: an object that moves and a constant force that pushes or pulls on the object in the direction in which the object moves. Physics uses the term "work" to denote the product of force times the distance over which the force acts:

$$\text{work} = \text{force} \times (\text{distance over which the force acts}).$$

In the example of the preceding paragraph, the force is first your upward pull and then your upward push on the suitcase, and the "distance over which the force acts" is the distance from floor to luggage rack. In agreement with commonsense, the larger the force, the more work done, and the greater the distance, the more work done.

Energy

In physics, *energy* means the ability to do work. Energy is an attribute (or "property") of a physical object or of whatever is contained in a specified region of space.

Forms of energy

Energy comes in several forms, three of which we note here.

(1) *Kinetic energy:* the energy associated with motion. The word "kinetic" comes from the Greek root *kinein*, "to move." The root appears also in the word "cinema," a name for the place where you go to watch "movies."

Here is an example of kinetic energy. A fast-moving hammer has energy of motion; when it strikes a nail, the hammer exerts a force through a distance, doing work as it drives the nail into the wood. Because the hammer, by virtue of its motion, has an ability to do work, we say the hammer has kinetic energy.

(2) *Potential energy:* energy that has the "potential" for being converted into kinetic energy.

Again, an example is in order. A sitting child, poised at the top of a slide, has (gravitational) potential energy. When the child lets go with her hands, she whooshes down the slide. The potential energy is being converted to energy of motion, that is, to kinetic energy.

(3) *Radiant energy:* the energy of electromagnetic waves.

For example, while lying on the beach, you receive radiant energy from the sun. An oven at 350 degrees Fahrenheit roasts a chicken with radiant energy, primarily infrared radiation, and a microwave oven does much the same, but with electromagnetic waves of much longer wavelength. In both ovens, the food absorbs electromagnetic radiation, thus absorbing energy and hence heating up.

The units of energy

Occasionally, it will be useful to know the units in which energy is expressed. Because energy is the ability to do work, it has the same units as work. In turn, the units of work are those of the product force times distance. The units of distance are easy: meters (in the metric system). But what about force?

Here it is worthwhile to return to Newton. He recognized that a force (acting in the absence of other forces) causes the velocity of a body to change. For example, imagine throwing a stone horizontally from a cliff overlooking a lake. You can watch as the earth's gravitational pull, acting on the stone, causes the velocity to tip downward (a change in direction) and to increase in magnitude. The rate at which velocity changes with time is called the *acceleration*. What Newton recognized, in short, is that a force (acting in isolation) produces an acceleration.

How much acceleration a specific force produces depends on the object. Some objects are very sluggish, very reluctant to change their state of motion. A cement truck stopped at a red light is a favorite example. Other objects, such as a ping pong ball or a fluff of dandelion, can easily be induced to change their velocity. In short, some objects have more inertia than others. The technical, quantitative term in physics for this property is *mass*. The mass of a body denotes its inertia, its sluggishness, its reluctance to undergo a change in velocity. Thus mass is an attribute of an object.

Newton's investigations – and ruminations – are summarized in the phrase "force equals mass times acceleration." A force produces an acceleration, but the more mass an object has – that is, the more inertia it has – the smaller the acceleration will be.

Now we have all the elements we need. The units of force must be those of mass times acceleration. The unit of mass is the kilogram (in the metric system). Because acceleration is the rate at which velocity changes with time, it must have the units of a change in velocity over a change in time. Thus acceleration must have the units of (meters per second)/seconds. Hence the units of force must be these:

$$\begin{pmatrix} \text{units of} \\ \text{force} \end{pmatrix} = \begin{pmatrix} \text{unit of} \\ \text{mass} \end{pmatrix} \times \begin{pmatrix} \text{units of} \\ \text{acceleration} \end{pmatrix} = \text{kilogram} \times \frac{\text{meters}/\text{second}}{\text{seconds}}.$$

Because the units of energy are those of work, we have

$$\begin{pmatrix} \text{units of} \\ \text{energy} \end{pmatrix} = \begin{pmatrix} \text{units of} \\ \text{force} \end{pmatrix} \times \begin{pmatrix} \text{unit of} \\ \text{distance} \end{pmatrix}$$

$$= \text{kilogram} \times \frac{\text{meters}/\text{second}}{\text{seconds}} \times \text{meter}$$

$$= \text{kilogram} \times (\text{meters}/\text{second})^2.$$

An amount of energy equal to 1 kilogram (meters/second)^2 is called one *joule*, named after James Prescott Joule. An Englishman, Joule studied the relationships among kinetic energy, potential energy, and heat in the middle of the nineteenth century, contributing greatly to what we now recognize as the law of conservation of energy. (On his honeymoon in Switzerland, Joule took along sensitive glass thermometers. He used them to measure the increase in water temperature after the potential energy at the top of a Swiss waterfall had been converted to large-scale kinetic energy during the plunge and then into microscopic kinetic energy when the water hit the pool at the bottom. On particularly rough mountain roads, Joule did not entrust the thermometers to the coach; he walked and carried them by hand.) The abbreviation for the joule is the capital letter J. For example, the energy that you can get by metabolizing 100 Calories of breakfast cereal is 4.82×10^5 joules or 4.82×10^5 J.

Conservation of energy

In newspapers and in the discourse of concerned citizens, one hears about "conservation of energy." The injunction "conserve energy" means "use certain forms of energy thoughtfully and as sparingly as possible." It is good advice.

The phrase "conservation of energy" has another meaning, too, a meaning that is prominent in physics considered as a pure science. Beginning with the investigations of Robert Mayer and James Prescott Joule in the mid-1800s, physics learned that energy, when considered in all its forms, is automatically conserved – in the particular sense that the total amount of energy remains constant in numerical magnitude. Energy may be changed from one form to another and then to a third form, for example, from gravitational potential energy (in water stored high above the turbines and generators that sit at the base of a dam) to electrical energy and then – in your toaster – to heat and radiant energy. As the amount of one form of energy decreases, the amount of another increases. The numerical changes compensate each other perfectly. Thus, if we take the numerical values of energy in all its manifold forms and add the values together, we find that the sum remains constant in time. Physics speaks of a *law of conservation of energy*. This is, of course, not a political law but rather a summary of one aspect of how nature behaves.

To be sure, now you may wonder, if energy is automatically conserved (according to physics), why should I make special efforts to conserve energy (in the environmental sense)? The answer is that some forms of energy are more useful to us than others, and the transformation of energy from the less useful forms to the more useful forms exacts a price. But to do justice to this process and its ramifications – the subject matter of thermodynamics – would take us too far from the trail of light.

Appendix B   The dependence of mass on speed

How does the mass of an object depend on its speed relative to us? In this appendix, we work out the answer. The key elements in the derivation are momentum conservation and time dilation. There is no need for a totally new idea; we just need to use familiar ideas in a carefully-constructed context.

We start off with two objects that have been built identically. We can visualize them as two iron spheres, each the size of a person's fist and containing identical numbers of iron atoms. One sphere is at rest in Alice's frame; the other, in Bob's. Figure B1 illustrates this.

Alice's frame moves with speed v relative to Bob's, and so Bob notes, as the initial situation, the configuration sketched on the top left in figure B2.

One item remains to be added to each iron sphere: a small pistol cap glued to the side that faces the other sphere. As the spheres pass each other, the slightest of contacts detonates the caps, and a tiny explosion pushes the spheres apart. The central and right-hand sketches in figure B2 show the explosion and its aftermath.

We focus our attention on the amount of momentum perpendicular to the

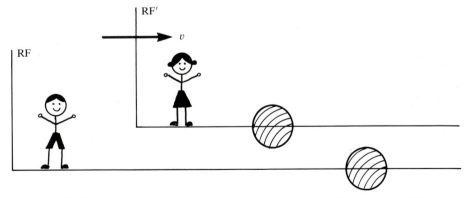

Figure B1 Two iron spheres in relative motion. Here the vertical displacement of Alice's frame is real. The spheres are *not* destined for a head-on collision; rather, they will barely make contact.

315

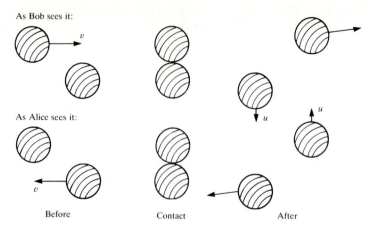

Figure B2 At the instant depicted in the central column, a tiny explosion occurs, pushing the two spheres apart in the vertical direction.

relative motion of the reference frames. Initially, the spheres had no motion in the perpendicular direction; so that component was zero to start with. Action *internal* to the physical system, such as the tiny explosion, cannot alter the momentum of the spheres taken in combination. So we must have zero momentum afterward in the perpendicular direction. In short, the downward momentum of Bob's sphere must be equal in magnitude to the upward component of the momentum of Alice's sphere. Here is where momentum conservation enters the analysis.

Next, we need to work out some velocity components. Let u denote the speed of Bob's sphere as observed by him. (We reserve the letter v for the relative motion of the frames.)

What does Alice observe for the speed of her recoiling sphere? From Alice's point of view, it was Bob's sphere that initially moved (with speed v to the left) and came into contact with her sphere, initially at rest in her frame. The relationships are symmetric, and so Alice observes her sphere to move solely in the perpendicular direction and with speed u, just as Bob does for his sphere. A tabulation is begun in table B1.

Now, what does Bob observe for the perpendicular velocity of Alice's sphere? We need to compare perpendicular distances traveled and times

Table B1. *Velocity components* perpendicular *to the relative motion.*

Bob observing his sphere	u
Alice observing her sphere	u
Bob observing Alice's sphere	$u \sqrt{1-(v^2/c^2)}$

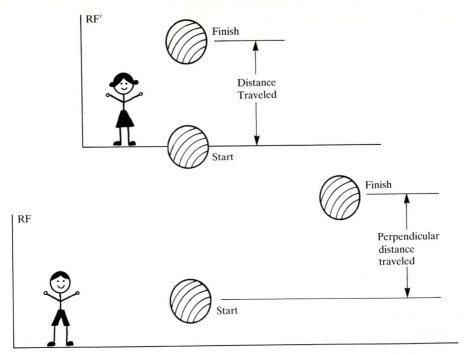

Figure B3 The motion of Alice's sphere, as noted by both Alice and Bob. (Bob's sphere is not shown.) The "start" event is the explosion event; the "finish" event is any convenient later event in the sphere's history.

elapsed. This is illustrated in figure B3. Precisely because the distance in question is perpendicular to the relative motion, Alice and Bob concur on the distance traveled. But they disagree on the elapsed time. Focus on events at the beginning and end of the speed measuring interval, as noted in Alice's frame. They are like the first event and the reflection event in the time dilation analysis of figures 10.1 and 10.2. There we found that the time interval Bob measures is longer than what Alice measures by the factor

$$\frac{1}{\sqrt{1-(v^2/c^2)}};$$

the same relationship is true here. The same geometrical arrangement of the events with respect to the relative motion implies the same time dilation factor.

Thus Bob observes the same perpendicular distance as Alice does but a time interval longer by the factor $1/\sqrt{1-(v^2/c^2)}$. That means Bob observes a slower speed. Specifically, Bob observes the perpendicular velocity component of Alice's sphere to be only $u\sqrt{1-(v^2/c^2)}$. This is duly entered in table B1.

We can return to momenta and concentrate on the way Bob observes conservation of momentum. Because the tiny explosion gives Bob's sphere only a very small speed (relative to c), the mass of Bob's sphere (as observed by him) is essentially its rest mass m_0. Its momentum is $m_0 u$. Alice's sphere, as observed by Bob, has mass m (a presently unknown quantity) and perpendicular speed $u\sqrt{1-(v^2/c^2)}$. The perpendicular component of momentum of Alice's sphere must equal that of Bob's sphere (in magnitude), and so we may write

$$mu \sqrt{1-(v^2/c^2)} = m_0 u.$$

Dividing both sides by u and by the square root gives the relation

$$m = \frac{m_0}{\sqrt{1-(v^2/c^2)}}. \tag{1}$$

On the left-hand side, the symbol m denotes the mass of Alice's sphere as observed by Bob. Alice's and Bob's spheres contain identical numbers of iron atoms and hence have identical rest masses. Thus we may take the symbol m_0 on the right-hand side to be the rest mass of Alice's sphere. Then equation (1) expresses the mass of Alice's sphere, as observed by Bob, in terms of the sphere's rest mass and its speed relative to Bob. Note that, because the divisor is less than 1, the mass m is larger than the rest mass m_0. Alice's sphere, as observed by Bob, is in motion and has kinetic energy; that energy contributes to the sphere's inertia.

How general is the result in equation (1)? Nothing in our analysis depended on the shape, size, or constitution of the objects. Iron spheres with caps just made the thought experiment easier to visualize. We may take the relationship expressed in equation (1) as valid for any physical object for which a rest mass m_0 has a meaning.

Appendix C   More about $E=mc^2$

In chapter 11, I noted that the equation $E=mc^2$ has given rise to a host of misconceptions. There are other issues associated with the equation, too, among them an issue of definition, and this appendix addresses some of those issues. It is written primarily for an instructor, but a student who has read appendix B will be able to understand most of it and will gain a greater perspective.

Mass in history

The word "mass" has a long history and a correspondingly long string of meanings and nuances. Max Jammer provides a fine discourse on the topic in his *Concepts of Mass in Classical and Modern Physics*, cited in the "Additional resources" for chapter 11; I rely on Professor Jammer for the early history. In Roman times, the Latin word "*massa*" denoted a lump of dough. This sense of "lump of stuff" is what we have in mind when we "hang a mass from a spring." By Newton's time, the word "*massa*" (still in Latin, of course) had come to denote "quantity of matter," a valuable concept but one that defied attempts to make it both operationally precise and independent of any prior notion of density.

Newton recognized clearly the physical notion of inertia and hence what we would today call "inertial mass." It was Leonhard Euler, however, who first defined inertial mass operationally as the ratio of force to acceleration (more precisely, as the ratio of their magnitudes). The constancy of the ratio (at least within the experience of seventeenth- and eighteenth-century physics) provided a characteristic value for each specific body: its inertial mass.

When Einstein was striving to develop his theory of gravity – the decade 1905–15 – physics recognized three distinct but related notions of "mass" for a specific body: its inertial mass, its active gravitational mass (meaning the extent to which the body produces a gravitational field), and its passive gravitational mass (meaning the extent to which the body is pulled on gravitationally by other bodies and hence a quantity proportional to the body's "weight"). Remarkably, all three of these "masses" are universally proportional to one another, a property recognized already by Newton.

To recapitulate, in the history of physics we have at least five distinct notions of "mass": lump of stuff, quantity of matter, inertial mass, active gravitational mass, and passive gravitational mass. (No wonder students are often confused.) For the narrow purpose of discussing the equation $E=mc^2$, however, we need only one notion of "mass": inertial mass. To put it positively, the "mass" that appears in $E=mc^2$ is the inertial mass. The history of the equation itself leaves no room for doubt on that score. (A good reference is Willard Fadner's article, cited in chapter 11.) Einstein and others spoke explicitly about "inertia" (the German word being "*Trägheit*") and "inertial mass" ("*träge Masse*"). Indeed, Einstein's first paper on the subject had for its title the question, "Does the inertia of a body depend on its energy content?"

Only semantics

Even after one has agreed that, in our context, "mass" means "inertial mass" and hence means "inertia," further specification is required. Let me set the scene with an example.

When an electron, say, moves at a substantial fraction of the speed of light, its momentum is related to its speed by the equation

$$\text{momentum} = m_0 \frac{1}{\sqrt{1-(v^2/c^2)}} v, \tag{C1}$$

where the symbol m_0 denotes the electron's inertia when it has zero speed. What should one call here the electron's "mass"? Three options arise:

(1) One can call the entire coefficient of v the "mass," in which case the constant m_0 acquires the name "rest mass."

(2) One can call m_0 the "mass" (leaving the square-root factor to convert simple v into part of the relativistic velocity four-vector).

(3) One can turn to an entirely different expression and use it to define what one will call the electron's "mass":

$$\text{"mass"} = [E^2 - (\text{momentum})^2 c^2]^{\frac{1}{2}} / c^2, \tag{C2}$$

where E denotes the electron's energy.

There is no consensus on which option to adopt, and my use of the verb "call" indicates that the issue is one of semantics. Each option has its advantages and disadvantages. So long as people are clear that they are just making a choice of definition, it does not matter which option they choose.

In the derivation in chapter 11, I adopted option (1) implicitly. It provides the greatest continuity with the momentum and velocity relation in Newtonian physics, and – as we saw – it leads to $E=mc^2$. A disadvantage is that one needs to be circumspect in relating mass to force and acceleration. An operational definition of mass as the ratio of force to acceleration (in magnitudes) requires the stipulation that the context be one of constant speed (as in magnetic deflection). This is reasonable, for if "mass" may depend on speed, then one ought to measure it in an experiment where the speed does not change.

In option (2), "mass" is independent of speed, and its numerical value is an intrinsic property of an electron, a nice advantage. The equation $E=(\text{mass})c^2$ holds, however, only for an electron at rest because the energy E varies with speed but this option's "mass" does not. (There is a whiff of inconsistency, too. The kinetic energy of the protons moving inside a nucleus contributes measurably to the mass of a stationary nucleus (according to all options). Shouldn't the kinetic energy of a proton outside a nucleus contribute to the proton's "mass," in which case "mass" would be dependent on speed?)

In option (3), the right-hand side of the defining relation is a relativistic invariant, that is, its numerical value is the same in all unaccelerated reference frames. That is a nice property. The property even enables us to evaluate the expression readily when the object is an electron. We choose to observe the electron in its own rest frame, where its momentum is zero and its energy is m_0c^2. All factors of c cancel, and so option (3) says that the electron's inertia when the electron has zero speed is the electron's "mass" in general.

One might now think that options (2) and (3) are equivalent, but in a vital sense they are not. When applied to an electron, options (2) and (3) do indeed have identical implications: an electron's "mass" is always just m_0. But option (3) can also be used for "objects" that are never at rest in any frame of reference, objects to which equation (C1) is not applicable. Most significantly, option (3) can be applied to a photon. For a photon, $E=hf$ and $(\text{momentum})=hf/c$, and so the squares in equation (C2) cancel each other perfectly. Thus option (3) says a photon has zero "mass." (Worth noting is that, even if one adopts option (3), the photons in a hot oven increase the oven's inertia (relative to the oven's inertia when cold). Although the photon has been assigned a zero "mass," photons do contribute positively to a system's measurable inertia. Moreover, Einstein's remark of 1905, "if the theory corresponds to the facts, then radiation carries inertia between the emitting and absorbing bodies," is entirely correct; the conclusion follows from the premise and is true experimentally.)

It is not my intention to advocate any one of the three options, but I had to make a choice for this book, and it was option (1). (By specifying slow speed for the atom as observed by Bob and by taking a limit of zero speed, one can convert the derivation in chapter 11 to option (2).) Here I offer merely a personal perspective on the options. For a physicist working regularly with four-dimensional space-time, options (2) and (3) seem preferable. For a physicist focussing on the deep connection between inertia and energy – notions that are both historic and contemporary – option (1) seems more desirable. Of course, no conclusion about physics depends on which option is chosen; only the words used to express the conclusion may differ.

The root of the linguistic problem seems to be this: each physicist would like the short word "mass" and the unadorned symbol m for the concept that he or she uses most often. The physicist wants to avoid phrases like "the relativistic mass" (for the coefficient of v in equation (C1)) or "the rest mass" (for m_0) or "the invariant mass" (for the right-hand side of equation (C2)). In speech and in writing, such phrases are cumbersome in comparison with "the mass." Until the day comes when unanimity of usage reigns, a reader will have to dig out an author's working definition and then read or translate accordingly.

Forms of energy

We say that water comes in three "forms": solid, liquid, and vapor, typified by an ice cube, what we swim in, and the steam in an old-fashioned steam locomotive. Here we have an *actual substance*, namely a collection of water molecules, that exists in different physical states.

What about the "forms" of energy? Because energy is "the ability to do work," energy is not a substance. There is always something out there in space that possesses the "ability," that possesses the attribute in some quantitative measure (such as 17 joules). That something may be a tennis ball or a charged battery or an electromagnetic field, to mention just three examples.

We often talk about energy as though it were a substance, but it is not. Rather, motion confers the ability to do work, and so kinetic energy arises as an attribute. Position can confer the ability to do work, and so potential energy arises as an attribute jointly of object and source of force (as in an apple and the earth or an electron and the nucleus of an atom). Electromagnetic radiation has the ability to do work, and so we say that electromagnetic waves possess radiant energy, again an attribute, just like the waves' momentum. What we seem to mean by the phrase "forms of energy" is that we can link the attribute "ability to do work" with other attributes of a physical system, such as motion or position or wave amplitude.

How can we be sure that energy is not a substance? Relativity theory itself provides reasons; perhaps one will suffice here, as follows.

To begin with, let me note that physicists take momentum to be an attribute, not a thing or a substance. Because momentum is a quantity with a direction as well as a magnitude, momentum has (in general) components along all three directions (x, y, and z) defined by the axes of a reference frame. Relativity theory collects an electron's energy and the three components of its momentum into a single quantity with four components. (The technical name for that entity is the electron's energy-momentum four-vector.) Because the theory places energy and momentum on the same footing in that quantity, energy and momentum had better have the same ontological status: they must exist in the same way. Since momentum is surely an attribute, not a substance, the same must be true for energy.

In short, it is proper to say that something *has* energy. It is *not* correct to say that something *is* energy.

Further defense against misconceptions

The primary defense against misconceptions lies in remembering that, in the physics of $E=mc^2$, "mass" means inertia. This provides us with a test: if we cannot sensibly substitute the word "inertia" for the word "mass" in an author's sentence, then the author is misusing the word "mass," and the entire sentence is deeply suspect.

For another element in the defense against misconceptions, we should bear in mind that neither inertial mass nor energy is a substance; rather, both are attributes of a physical system. "Matter," however, is most certainly a thing. An attribute cannot be converted into a thing, and so energy cannot be converted into matter. Likewise, a thing cannot be converted into an attribute, and so matter cannot be converted into energy. To be sure, these statements are nothing more than elaborations on the theme of permissible conversions discussed in chapter 11.

The Great Heresy

In chapter 11, I tried hard to make only correct statements, but in teaching a class about $E=mc^2$, one may need to make a false statement and then explain why the statement is false. This tactic can expose a deeply ingrained misconception. The false statement that I have in mind, which I call *The Great Heresy*, is this:

> The equation $E=mc^2$ means that you can convert mass into energy and vice versa.

If we try the substitution test, replacing the word "mass" with the word "inertia," then our ears suggest that the claim cannot possibly be correct. The fallacy in The Great Heresy arises from a simple confusion. A glance back at figure 11.5 puts the essential relationships before us. The heresy confuses the convertibility of matter and radiation with the parallel changes of energy and inertia. The heretic takes the word "mass" to mean "matter," as in the case of an electron–positron pair. Moreover, the heretic takes energy to be an entity existing in its own right, rather than an attribute of some physical system, such as electromagnetic radiation. It *is* true that "you can convert matter into radiation and vice versa," but that truth is *not* what the equation $E=mc^2$ means.

As analyzed here, The Great Heresy confuses two relationships. To me, the convertibility of matter and radiation is the more surprising feature of nature. No wonder, I say to myself, that it has gotten great play. But that convertibility is distinct from the implications of $\Delta E=(\Delta m)c^2$: as the energy of a system grows, so does its inertia. And then the equation $E=mc^2$ tells us that, for any physical system, its energy E is always accompanied by an inertia whose numerical value is given by the expression E/c^2.

Another way in which The Great Heresy may arise – or may be construed – appears in the context of nuclear fission, to which we now turn.

Uranium fission

The law of conservation of energy is all that one needs in order to understand the "energy release" in nuclear fission. A nucleus of uranium-235 absorbs a slow neutron, say, and undergoes fission. After the splitting, the electrical repulsion between the two large fragments imparts tremendous kinetic energy to the fragments as they accelerate away from each other. A few fast neutrons are emitted – that's more kinetic energy – and there is some prompt gamma radiation. Since some forms of energy have increased in value, others must have decreased. Indeed, the internal potential energy – electric plus nuclear – of the fragments is less than that of the original nucleus. The fission process – in which the number of protons and neutrons does not change – is primarily a transfer of energy from potential form to kinetic.

The Great Heresy may be construed as meaning that "rest mass" is converted into energy. "Rest mass," however, just means "intrinsic inertia," and that inertia is a property of the energy inherent in an object. In fission, the sum of the final rest masses is less than the sum of the initial rest masses because of some quite ordinary changes in the form in which energy is present. After fission, there is less potential energy but more kinetic energy and the energy of radiation. Because the final aggregations of protons and neutrons have less potential energy in them than did the original aggregations, they have less inertia if put at rest than did the original nucleus and neutron. So, of course, there is a reduction in the sum of the rest masses. That reduction is *not* the *source* of the energy released; rather, the reduction is just a concomitant of the release.

To be sure, the numerical value of the reduction in rest masses can be used to calculate the amount of energy released as kinetic energy and the energy of radiation. Section 11.6 shows how to do that. The essence of the calculation is merely $\Delta E = (\Delta m)c^2$ applied to the potential energy in the aggregations and to the inertia of those aggregations. To learn how much the potential energy has decreased, one looks to see how much the inertia has decreased.

Why introduce anti-particles?

In the fall semester of 1990, when I had finished deriving the equation $E = mc^2$ in the fashion of section 11.5, one of my students asked, why do we need to introduce anti-particles and the creation of an object? My answer was this: so that we can start with zero for both the object's energy and its inertia. *Without* those two zeroes, the equation $\Delta E = (\Delta m)c^2$, when applied to increments, would imply merely that "$E = mc^2 + \text{constant}$," and we would not know what value the constant has. To establish a zero value for the constant, we need to start from zero for both E and m.

The anti-particles provide a clean derivation. Moreover, their annihilation with particles (in a separate process) makes clear the distinction between the convertibility of matter and radiation and the parallel changes of energy and inertia.

Index

acceleration, defined, 306, 312
alpha decay, 253
alpha particle, 125–6
amplitude, defined, 66, 306
angle:
  critical, 10
  of reflection, 3
  of refraction, 3
anti-particles, 242–7, 325
atom, defined, 306

beta decay, 254
beta ray, 125–6, 254

color:
  Newton's particle theory and, 40
  Snell's law and, 17–18
  wavelength and, 91–3
  white light and, 16
complementarity, 170, 174
component, defined, 37, 306
Compton effect, 148–54
Compton, Arthur Holly, 149–54, 161
coordinates of event, defined, 285–6

$E = mc^2$:
  atomic bomb and, 248, 259–66
  conversions and, 242–5, 251–2
  derivation of, 238–42 plus 246–8
  meaning of, 247–8, 266–9
  misconceptions about, 248, 268–9, 323–5
  parallel changes and, 244–5
  rest mass and, 250, 268
  scales of energy release and, 256–60
  The Great Heresy and, 324
Einstein, Albert:
  atomic bomb and, 262–4
  biographical sketch of, 195–206
  $E=mc^2$ and, 244, 247, 259, 269
  Nobel Prize for, 154–8
  photon and, 147, 155–6, 158, 173
  special theory of relativity and, 185, 199–201

electric current, defined, 114–15
electromagnetic waves (see waves)
electromagnetism and technology, 133–4
electron:
  contrasted with photon, 171
  defined, 106, 306
energy:
  as attribute, 238, 322–3
  conservation of, 145, 151, 237, 313–14
  defined, 236, 306, 311
  forms of, 236–7, 311, 322–3
  kinetic, defined, 236, 307, 311
  potential, defined, 236, 308, 311
  radiant, defined, 311
  rest mass and, 250, 268
  special theory of relativity and, 323
  units of, 311–13
event, defined, 193, 306

Faraday's law, 116–18
field:
  electric, defined, 108–9
  magnetic, defined, 113
  why introduced, 110
fission:
  described, 248–53
  Hahn *et al.* and nuclear, 254–5
  in uranium, 248–9, 254–5, 324–5
  induced, 248–9, 252–3
  spontaneous, 252–3
force, defined, 35, 306, 310
Foucault, Léon, 44–5
frequency, defined, 64, 306
Fresnel, Augustin, 95, 101–3
fusion, 253

gamma ray:
  conversions with matter, 242–4
  defined, 125–6, 252
  in nuclear processes, 248–9, 252, 256

half life, defined, 219, 306